Raising Chickens

Raising Chickens

The essential guide to choosing and keeping happy, healthy hens

Suzie Baldwin

Photography by Cristian Barnett
Illustrations by Becca Thorne

KYLE BOOKS

Contents

My now intimate relationship with chickens started shortly after I purchased my house in the country. Little did I know how close we would become and how fond I would grow of them.

When my first chickens arrived we housed them in temporary accommodation which was an old shed/summer house cordoned off with wire. When I was home I would let them out during the day to roam the garden. The chickens became so tame they would wander over to the kitchen door to say hello – perhaps with ulterior motives in the snack department. At sundown I would call and they would come running, following me right back into their hut. My friend Shirley and her dog Maisie often came to stay and the chickens would also quite happily follow them around the garden with Maisie on a lead. Then came the foxes and a more secure residence was built which a friend called 'Peckingham Palace'.

I have had my current set of three lovely Light Sussex for two years now. They produce the best eggs in the world. The best I have tasted anyway; creamy and delectable. Far superior to shop-bought eggs. People ask me how I cook my eggs and I have to explain it is the eggs themselves that are unmatchable.

When working away from home I miss keeping my 'scraps' of food that I put aside for the chickens; I hate waste. My sister Natasha had chickens before me on her farm in New York and she swore that their preferred meal was pasta and salad.

Earlier this spring one of my chickens was poorly. Some of her feathers were missing and her comb was pale. I talked to my very knowledgeable friends at the farm where I bought my hens and they advised putting vitamins in the water. This pink coloured tonic worked a treat. Two weeks later her comb was a vibrant red again. I called my mother in America prattling on about how happy it made me to have nursed our chickens back to health.

Take my word for it, you become involved and fascinated. Sometimes with all the travelling I do I wonder why I keep hens. However the moment I get home I wander out into the garden to collect the newly-laid eggs. They are a pale brown colour, but worth their weight in gold.

Enjoy.

Joely

Why keep chickens?

Keeping chickens

When I started keeping chickens, more than twenty years ago, it was looked upon as an odd, even bizarre, thing to do and I was regarded as madly eccentric! Today, thankfully, keeping chickens as pets is an increasingly popular hobby, one that people from all walks of life, young and old, are realising they can do, with the result that it is considered almost normal.

Chickens, if kept properly, are very rewarding – and very addictive. Once you have had even one or two it is difficult to live without them. Many people who have lost their girls through old age or the dreaded Mr Fox say it leaves a huge hole in their lives, so be warned. A close friend who lost her chickens to a fox told me she was so heartbroken she would leave it a while before replacing them. I was surprised, therefore, to see her a week later, box in hand, desperate for some more. Without chickens, she said, her garden had lost its soul, and the slugs had taken over the flowerbeds, plus she had put on a whole pound in weight as there were no girls to share her morning biscuit with, and the phone bill was enormous as there were no girls to talk to.

There are a few things to consider before you buy your chickens: are there any local regulations, by-laws or clauses in the deeds of your property that restrict you from keeping poultry? Do you have the time and resources? Chickens are cheap to look after but the set-up cost can be relatively expensive and they need protecting from predators. You need to be able to spend time with your girls to benefit not just from their eggs but also their enchanting characters. Think about the noise. Chickens are relatively quiet, providing you have no cockerels, and they are at their noisiest when telling everybody they have laid an egg. You will also need to provide a chicken sitter for your girls when you are away on holiday. You can normally persuade a neighbour or friend with the bribe of letting them keep the lovely fresh eggs.

If you grow your own veg, chickens do a fabulous job of not just fertilising your garden, but scratching up weeds and generally cleaning it for you. They can attract rodents but if you keep them clean and are careful with their feed, you should be fine. If you are a keen gardener and love your flowers, do not let the chickens near them – their idea of a garden and yours are totally different! I describe my girls as my walking flowers.

If you feel that chickens are too much of a responsibility for you, why not share them, and the chores, with friends or neighbours? If you have a large garden or field, you could provide the land. More and more chicken syndicates are being set up, some on allotments, others on private sites.

RIGHT: A friend's thoroughly spoilt Lemon Millefleur Booted Bantam hen.

In my experience, if you choose to keep chickens your life will change for the better. They are great stress reducers and endlessly fascinating – you can watch them for hours. It's also lovely to come home and be greeted by energetic clucks after a hard day at work, plus many people find they make a new network of friends who are happy to offer help and advice and also buy their eggs. In addition, chickens teach children responsibility and respect. They soon learn that if they look after their chickens well they will reap the rewards in eggs, in the satisfaction of a job well done, and also, believe it or not, in the acquisition of business skills. I have known several children who have bought chickens and gone on to earn money selling their eggs.

My aim in this book is not only to give you a thorough guide to chicken keeping but also the confidence to happily look after your charges and avoid common problems. As with all animals there can be difficulties, but the way you set up and treat your girls from day one will determine how successful and enjoyable your experience is. Most chicken problems can be prevented and the more information you have to hand before you start, the less stressful and scary it will be – both for you and your chickens. With sections on everything from what to look for before buying chickens; getting your chickens to go to bed on their first night in their new home; maximising your egg yield; dealing with broody chickens; ensuring your chickens are happy, whatever the season; keeping them healthy; and what to do when they reach the end of their lives; this book will help everyone from the complete beginner to the more experienced chicken keeper to get the most enjoyment from their flock.

In my opinion, chickens are easier to keep than dogs – they don't need taking for walks – and they fit in very well with most lifestyles. They will soon recognise your voice and come running when they hear you, and give you more than love – fresh eggs! With care, and by following helpful tips based on years of experience, you can have calm, manageable egg layers eating out of your hand.

Keeping chickens has changed my life. I have made many friends on the journey, both young and old, and had the pleasure of collecting fresh eggs, which is especially pleasurable on a cold day when they make great hand warmers.

COURSES ON CHICKEN KEEPING

Many chicken suppliers run courses that are hugely popular and which past students have told me are very beneficial. Classes can be both informative and hands on, which is great if you are lacking in confidence. Also you get the chance to meet like-minded people, which can lead to new friendships. I know of one course where three ladies really hit it off. The confidence they gave each other was fantastic, and they are still in regular contact. Check the internet to see if there are any courses running near you. Having attended one, you will know if chicken keeping really is for you.

RIGHT: A free snack of freshly picked blackberries being relished by my Buff Orpingtons.

LEFT: You can never spend too much time watching your girls.

FIVE REASONS TO KEEP CHICKENS

1 A supply of fresh, free-range eggs – every day. Most 'fresh' eggs bought from a supermarket are at least three weeks old. Research has also shown that organic, free-range eggs are much lower in cholesterol than eggs that come from intensive production systems.

2 They are the very best way of instigating a sense of care and responsibility in children. Children also love to collect eggs and doing so helps to connect them with the reality of how food is produced.

3 They are easy pets to look after and often pay for themselves. There is also a breed out there to suit everyone and almost all gardens.

4 For their entertainment value. Chickens are endlessly fascinating and simply being near them gladdens the heart and lifts the spirits.

5 Chicken droppings are rich in nitrogen, phosphates and potash and are a great fertiliser for the garden. Don't, though, put them directly on to your flowers as it will scorch them. Chickens also love to eat slugs.

FIVE COMMONLY ASKED QUESTIONS

1 How many chickens do I need? Depending on the breed, a hen will lay up to 300 eggs a year, so almost one a day (see pages 43–52 and page 105). I would recommend a minimum of three birds together because any fewer will bicker and quarrel.

2 Do I need a cockerel? No. Chickens like to have a cockerel around, but it doesn't affect how they lay. If you live in a built-up area it's probably kinder to your neighbours not to have one.

3 Where should I buy my chickens from? Always buy from a reputable supplier who is willing to give advice freely and offer help.

4 How long do hens live? A happy, free-range hen can live for anything from five to ten years, depending on the breed.

5 Do I need permission to keep chickens? If you are keeping chickens just for family use, you are unlikely to encounter any restrictions, but it is always advisable to check local planning regulations.

Housing, equipment and feed

Housing and equipment

To set up with your chickens you will need: a coop, a run, feeders and drinkers, bedding, feed and mixed grit, plus a dustpan and brush, paint scraper, disinfectant and gloves to keep them clean. And a basket to collect your eggs in.

A lady once visited my farm wanting to purchase two hens for her kitchen. When I said, 'You mean the garden,' she replied, 'No, I intend to keep them in a dog cage in the kitchen as I only want them for eggs.' I declined to sell her any of my girls!

TYPES OF HEN HOUSES

There is an abundance of chicken coops on the market to choose from. Avoid those with felt roofs as they harbour red mites and make it almost impossible to eradicate them, as treatments are unable to penetrate the felt. If a house is very cheap it usually means that it isn't built to last. You need to make sure that the wood has been treated and is of a decent thickness for insulation and to keep predators out. The locks and hinges need to be robust and of a good quality, as does the wire for the run. Ideally perches should be removable, as this makes them easier to clean. Nest boxes should be accessible for you to collect the eggs. A rule of thumb is one nest box for three hens. Decide if you want a fixed coop or a movable one. If it is movable you will be able to rest the ground regularly, although this does leave unsightly, worn areas. Fixed or non movable coops can be made fox-proof and customised for your needs.

You can also buy plastic houses which are well designed and apparently completely fox-proof. These have less chance of getting red mite as they have fewer crevices, although in my experience red mites can get in anywhere. If a plastic house does get red mites it is very easy to get rid of them by hosing out the coop, but this will not kill them as they will soon dry out and crawl back in.

You can also pick up second-hand coops, as people tend to progress from the small units to bigger ones as they find they enjoy chicken keeping and decide to increase their flock. The small plastic houses make interacting with your girls rather difficult, as you have to do everything on your knees.

I am definitely a wooden coop girl as I find them more natural and aesthetically pleasing.

RIGHT: Hybrid hens looking unsure in their new home.

HOUSE CHECKLIST

Timber: Make sure this is properly treated so it will last and withstand the weather.

Roof: Choose wood, not felt, with an overhang for shedding water. You may also be able to find a corrugated roof, which will give better ventilation, made from recycled fibres that have been dipped in bitumen to waterproof it.

Ventilation: Holes covered with mesh will provide fresh air without draughts – the number required will depend on the size of the coop.

Pop holes: The entrance for the chickens is normally 30cm by 30cm, usually with a sliding shutter or drop-down ramp with secure fastenings to keep out predators.

My favourite type of coop is a raised one with a raised run. It is easy to clean as it's at waist height, it provides a sheltered area underneath the coop to hang goodies, feeders and drinkers, and it deters rodents as there are no dark areas at ground level for them to rest in. The height of the run makes it feel airy and bright, even with shade on the roof, it lends well to having extension runs made easily, and provides different levels for the chickens to be in. It can usually be moved by two people, but can also be used in a permanent position.

How much space do you need?

The minimum legal requirement is 0.09 metre square per bird in the coop and 1 metre square for the run. However, I cannot emphasise enough that the more outdoor space you can give them, the happier they will be.

Before buying your house and run it is a good idea to think it through thoroughly and not jump in on a whim. Spend time checking out all the options, visit places that sell them and have a look around, ask questions and think outside the box! A customer came to see me and explained that she lives by a canal and has frequently seen a canal boat passing by with a chicken coop and run on the top. She felt that if you can keep chickens on a canal boat, she could certainly keep them in her garden.

Most gardens are suitable for chickens as long as you are sensible and do not keep too many. If you have a lot in an inadequate space, you are asking for trouble. Not only will you have to clean constantly, but you'll have a constant battle with bad behaviour and disease.

If you are on a budget you can often buy second-hand coops but make sure you check them over properly before buying, as it is so easy to take red mites home with you. A lady who was very upset once came into my yard. In the boot of her car she had a bargain – a second-hand coop and, free of charge, millions of red mites all over the car and herself.

You can also convert sheds or old Wendy houses, provided they have ventilation, perches and nest boxes. Ventilation can be made by drilling some holes above the door and covering them with wire to prevent predators getting in. Perches can be installed and nest boxes made very easily.

Lots of people come to my farm with pictures and videos of their coops and enclosures. Some are amazing – one couple had converted their summer house and it had double glazing and central heating, and its own letter box!

I have seen coops that have been painted to match the owners' house, brilliant coops made of wardrobes, wooden packing boxes, even a bath tub. Bearing in mind a chicken's requirements, let your imagination go wild.

Perches: These should be removable for cleaning, with a width of about 4cm and rounded for comfort. They should be placed higher than the nest boxes. If you have more than one perch, they should be arranged so the birds are not directly below each other, so they don't peck or defecate on each other.

Next boxes: The rule of thumb is one box for every three hens. Place them low down in the darkest part of the house.

Flooring: There is a choice of removable/slide out, fixed, slatted or wired for droppings to fall through.

Runs: These should be as big as the space allows, made with good-quality wire and covered with a wire roof.

TOP LEFT: A fully enclosed, walk-in chicken coop. TOP RIGHT: Collecting eggs. BOTTOM LEFT: Treats are always welcome.

Protection from the elements

LEFT: Shade is a must in very hot weather and can be as simple as adding a parasol.

Partial shade is essential – if the chickens are in full sun all day they can suffer from heat stroke. Shade can be provided by trees or shrubs, or a cover attached to one end of the run. Even a sun umbrella will work, and it can look really pretty.

Protection from the prevailing wind is important as you don't want it blowing straight into their house via the pop hole, or for the chickens to be blown over in the run. Place the coop near a wall or hedge, or you can use windbreakers which work well and are cheap and easy to use. It is also very important to choose a well-drained area for the run, as chickens don't like being covered in mud and puddles can breed disease.

Building your own hen house

Building your own coop is likely to save more than half the cost of a ready-made one, so it may be something you want to consider doing if money is an issue. The easiest type to build is an ark, which has a nesting space at one end and a run at the other, although a straightforward, rectangular hen house is almost as simple. The internet is a good source of ideas and information, plus plans, but make sure that whatever you build is large and comfortable enough for your chickens to rest at night and also to extend their wings during the day. You also need to ensure that in addition to a private and secure nesting area, your coop has a predator-proof run, easy access for cleaning and a way of letting your chickens in and out easily.

RUNS

It is easier and cheaper to buy unless you are very good at DIY. I make mine using two metre high fence posts, which are hammered into the ground, and attaching chicken wire with an overlap at both the top and bottom. The overlap at the bottom is important because it stops foxes from digging underneath. It needs to be secured with large tent pegs, concrete slabs or large logs. Logs provide a lovely habitat for bugs and frogs but if you use too many they'll also attract mice and rats. You only need a few to keep the wire firmly down. The overhang at the top is to prevent the fox from climbing in as it will wobble and make him fall off.

I have also used barbed wire all the way around the posts, both at the top and again at fox's nose level (about 15cm above the ground), but this is not a good idea if you have children. My sister managed to obtain some galvanised screens used on building sites, which work really well, but you do have to put a roof on top. She used plastic corrugated sheets which you can get from most DIY stores. The bottom also needs securing with either wire or concrete slabs to prevent anything digging in. Ground-level coops are good in small gardens as they are easy to move around, but they are vulnerable from predators around the bottom.

Chickens love to dust bath. It keeps them healthy, cleaning their feathers, removing excess oil and suffocating any mites that might be living on them. If they have a dry, sheltered area in their run with dirt in, they will make their own. If they are out every day in your garden they will soon find a favourite spot to dust bath, usually in a sheltered position under a bush, away from predators, so they can relax. If they have none of the above you can provide them with a dust bath very easily; all you need is a high-sided container – large seed trays filled with dry soil, sand ash or peat work well. You can also add red mite powder to their dry dust bath for both cleaning and prevention.

ABOVE: An overhang of chicken wire with barbed wire helps keep out Mr Fox. RIGHT: Houses can benefit from being in a sheltered area during the winter months.

POP HOLES

I have never used automatic pop holes as I'm a great believer in having contact with your chickens and these make it all too easy for you to not see them. Watching the chickens as you let them out in the morning and put them to bed is the perfect time to check on them. Having kept chickens for so long I am also well aware that occasionally, particularly on lovely summer nights, the girls will wander around the house rather than going inside. If you use an automatic closing device it is so easy for one or two to be locked out for the night as it works by sensing light and is incapable of counting your chickens in. A lot of my customers love these devices and I have heard many great reports about them. However, I cannot advocate their use.

Rats dislike the smell of mint, so if you have a problem with them try growing some near your run to deter them.

Electric fencing

LEFT: My girls free ranging in an electric fence enjoying a daily scratch of corn.

Electric fencing is a very good way of keeping your girls safe if used correctly. It is a flexible option as it is easily moved, allowing you to utilise different areas of your garden. But it does require maintenance, regular cutting of the grass around the fence and checking with a tester daily to ensure it is working. Some units have alarms that let you know when the power is low. I have used electric fencing for many years with great success, except for this year when a determined fox jumped in. Unfortunately once a fox has found a way in he will return, which meant I had to move my remaining chickens to a permanent fixed enclosure while I dealt with him. Since the fox has been dealt with I have had no further problems and am using the electric fencing again.

You do need to keep the power on for 24 hours a day, as when it is off it can be chewed by rodents, rabbits and foxes, which will cause damage and affect the voltage.

The bottom strand of the fencing is not electrified but all the others are, so if the grass or foliage are too high and touch the second line it will 'ground' or drain the power, making it less effective. Regular checking for any debris such as fallen branches is important, as this can also ground the power. Frogs and toads can also get zapped and die on the wire. This will affect the power, so a regular walk around when you let the chickens out and put them to bed is advisable.

There are many units available. Some run on mains power, which I prefer, but this does limit where you can use them. Batteries are cheaper and easily available, and some chargers work off solar panels and can be very effective. Whatever kind of charger you choose, it is best to buy direct from the supplier as they are usually on hand to offer help and advice. If buying from a retailer, ask about what telephone support you can expect.

RIGHT: A Poland cock nesting in straw.

BEDDING

The bedding you choose really depends on the weather. In the warmer months, sawdust is great but it needs to have the dust extracted as chickens are susceptible to respiratory problems. Horse bedding works really well and is much larger than normal sawdust. In colder months, straw is great as it also insulates the coop. Avoid using it, however, in the warmer months, as it provides a hiding place for mites (see page 124) and your hens do not need the added warmth. During the colder months, mites tend to be less active. Newspaper can be used, but be aware that some chickens use it as a toy. I once purchased two new layer houses that could take 50 chickens each and meticulously lined them with newspaper with a small amount of shavings on top. The girls loved it, but when I returned to check on them later they had removed the sheets of newspaper and were running around the field with them. The mess was horrendous but they were having a great game! Never use hay as this can be eaten by the chickens and form a tight ball in their crop, causing problems such as crop impaction (see page 128).

ENTERTAINING YOUR CHICKENS

RIGHT: Scarecrows are easy to make and can deter foxes, as can a radio left playing.

During the hot, dry months of summer my family and I spend a lot of time with our chickens. They wander around the garden scratching and bathing while we attend to the veg patch or flower beds, and there are plenty of bugs for them to chase. Life is great! It all changes as the nights get darker and the days colder. We tend not to be in the garden as much, so the chickens spend more time confined to the run. This is when your chickens could do with some entertainment. I don't mean a television set, although on occasions I have been known to pop a radio in the run for them. This provides different sounds that the fox doesn't like and the hens seem happy. Old CDs hung up give them something different to look at. A perch hung like a swing provides hours of amusement. Old logs for them to jump on and peck at are good too. You can fill the clear balls that have toys in (the type you get from vending machines in supermarkets), with mealworms and make some holes in them, so that when the chickens peck and push them around, the mealworms drop out. This is much more fun than just throwing the mealworms on the floor.

Chickens are very inquisitive and enjoy variety in their surroundings, so try different things and experiment – just make sure whatever you use is safe, with no sharp edges or small pieces that the chickens would be able to swallow. I enjoyed making a scarecrow which sat on a chair in the run. It didn't take the girls long to sit on its knee and explore it, and was something to make Mr Fox wary!

Feed

FEED

Feeding chickens is very easy nowadays; there is a huge choice of ready-made complete food on the market. Which kind you choose is entirely up to you, but do try to buy food that is as natural as possible and GM free. Some feed has colourants added which make the eggs yolks bright yellow.

Chicks from day one to about six to eight weeks need chick crumb. They then need rearer's or grower's pellets until they're 16 weeks old, and from then for the rest of their lives they need layer's pellets or mash. Layer's pellets are pressed into small, cylinder-shaped pellets, making it easier for the chickens to eat. Mash has the same ingredients but is a fine dust, which makes eating more time consuming for the chickens. This is seen as a benefit by some poultry keepers as it can help to occupy the chickens, giving them less time to become bored. However, I always feed my girls pellets. I find them cleaner and less wasteful. Chickens will waste a certain quantity of the mash and I also find that it becomes damp and clogs feeders as it draws moisture from the air.

Whatever you choose, it is always advisable when changing the feed to do it gradually, mixing half and half for a week or so until the chickens gradually become used to it. I believe in feeding layer's pellets on an ad hoc basis, meaning that it is available to the girls all day long.

The feed requirement of your chickens will vary depending on the weather; in colder weather they will consume much more in order to keep warm.

Never mix layer's pellets and corn together. Mixed corn is like chocolate to chickens and they will favour this over the layer's pellets, which contain the most goodness. But it is a great afternoon treat for them, best given by throwing a small handful of the corn into the run, allowing the chickens to scratch and forage for it.

WHERE TO STORE YOUR FEED

Most chicken feed – layer's pellets, mash and mixed corn – comes in 20kg bags. It is the cheapest way to buy your feed but needs to be stored in a cool, dry, ventilated place to prevent it from going mouldy. You should not store it directly on the floor but raise it using a wooden pallet or something similar. It is very important not to encourage any predatory local wildlife, so storage

WHAT TO FEED YOUR CHICKENS

This varies depending on the age of your chickens, as follows:

Hatch to 6–8 weeks: Chick crumb
6–8 to 16 weeks: Rearer's or grower's pellets
16 weeks on: Layer's pellets or mash

When changing feed do it gradually, mixing half and half for a week until your chickens become used to it.

TOP LEFT: Layer's pellets provide everything a girl could need. TOP RIGHT: Dandelions are loved by chickens. BOTTOM LEFT: Dock leaves are great to hang as a treat. BOTTOM RIGHT: Mixed corn is a great afternoon treat.

bins, either metal dustbins or plastic boxes with lids, work very well. If using plastic, be aware that vermin can chew through it, but not usually overnight, so check every day and if you see any signs take appropriate action.

Remember that the bags of food you buy have a 'best before' date. Although the food will look fine after this date, using it means that the nutrients and vitamins will have depleted. Most feed seems to have a shelf life of three months.

RIGHT: My girls enjoying a fresh harvest of hedgerow blackberries.

In the past, farmers fed their chickens swedes because they believed it increased the size of their eggs.

CHICKENS' FIVE-A-DAY

Chickens benefit from having a daily portion of green vegetables as they contain minerals. This is the equivalent of our five-a-day, and helps the chickens' digestive system to function well. Raw is best, and it is advantageous to hang the greens so the girls can peck at them more naturally. But too much cabbage and broccoli can cause diarrhoea, so just give a little of this.

Although lettuce has very little nutritional value it is great at keeping a hen's digestive system working well. It is also easy to grow yourself. Onions also aid the digestion and can help ward off many illnesses, but don't give them too many as they will make the eggs taste odd. Potatoes skins/peelings should be blanched before feeding. They love these in the winter months mixed with bran. Nettles are very good for them and free. Collect them when they are young, cut them up, boil them and add to their layer's pellets, or alternatively just hang them in bunches. Don't forget to wear gloves.

Other hedgerow treats your chickens will enjoy include blackberries, dandelions, dock leaves and chickweed, to name just a few. Always remove any uneaten food and discard it at the end of every day.

FEEDERS

There are many feeders on the market:

• **Treadle feeders** work by means of a paddle that the chicken steps on to open it. They are very good and prevent rodents being able to eat any food, but you will need to train your chickens to use them.

• **Outdoor galvanised feeders** are very robust and look attractive. They really need to be placed on a brick or level log as they are rather low for the girls if put on the floor.

• **Trough feeders** are quite wasteful as the chickens can scratch at them and also contaminate the feed by defecating in them.

• **Plastic feeders** that can be hung up are by far my favourite. You can hang them about 15 centimetres off the ground, which prevents the birds from scratching the food out of the feeder and from scratching mud and dirt into it. They are easy to fill and the food never becomes stale as it naturally filters out. They are also very easy to clean.

Whichever style you choose, you need to make sure you have sufficient space around the feeders for all the chickens to be able to feed at the same time.

DRINKERS

Chickens drink a lot and enjoy fresh, cool water. There are a lot of drinkers to choose from – the size you need will depend on how many chickens you are keeping.

• **Bucket drinkers** look great and are easy to use. They are galvanised buckets with half a lid. You just fill as normal and lay them on their side.

• **Jam jar drinkers** are easy to fill and use, and they can be hung up. They are old-fashioned and, being made of glass, look attractive but you need to be aware that they will break when dropped.

• **Plastic drinkers** aren't as pretty as glass ones but are robust and easy to use and clean.

• **Nipple drinkers** are very hygienic and easy to set up, and have the advantage of keeping the chickens busy.

Remember that many things can be used to hold water, but they need to be kept clean and hygienic. You don't want the poor girls cliimbing into them and you ideally want mud and dirt kept out. I've even seen discarded four-litre milk cartons used; they are actually very effective and free. Remember, too, that drinkers should be placed in the run and never in the house, as a damp bed area is not healthy for the chickens.

MIXED GRIT

Mixed grit, which contains flint and oyster shell, is an essential part of a chicken's diet. As they have no teeth, the flint is used to help grind the food in the gizzard. The oyster shell is a source of calcium and helps produce strong egg shells. There are grit stations on the market. They work very well but if you have a small run they can take up precious space. I find throwing a small handful of grit into the run regularly is fine. It encourages the chickens to forage naturally and doesn't clutter the run with another feeder. Grit doesn't attract pests, so you don't have to worry about that.

TOP LEFT: A selection of drinkers. TOP RIGHT: A galvanised bucket drinker. BOTTOM: A selection of feeders.

TREATS

Everybody deserves a treat occasionally and chickens are no exception. You will get as much enjoyment watching as they will get from eating them. Bake some cookies, but leave out the sugar and chocolate chips and add peanuts or raisins instead. Thread them on string and hang them up for the chickens to peck at. Fallen leaves collected and placed in a big pile will make them go crazy. Cereals threaded on to string and hung in the run provides them with hours of activity.

A smallish football placed in the run will be investigated for quite a while, and a treat ball that you can get from pet shops, filled with mealworms, will be pushed around for hours. Yogurt pots filled with treats and hung upside down will also provide entertainment. And corn on the cob hung up for them will make them jump for joy! Logs with holes randomly drilled in and stuffed with rice or peanut butter causes great excitement.

Old baby toys, rattles and mirrors hung up will make the chickens very inquisitive. Baby gyms are great; they also use them to perch on. Make fat balls for them filled with mealworms, sunflower seeds, stale breadcrumbs, etc. Grow sunflowers and collect their seeds in the summer to use as winter treats – they will help their feathers to grow.

Chickens are omnivores; they eat both vegetables and meat.

Chilled watermelons are very refreshing and help to keep chickens hydrated in the summer. Grapes (seedless) are loved if just thrown into the run. This is entertaining to watch as the girls get very protective of their find. You can buy live crickets from pet shops to give them, which makes a great game, and they are packed with protein.

Use bird feeders to stuff vegetables or fruit inside to make life a little more interesting. Threading cooked pasta on to string for the hens is something my children love to do and is always relished. Chickens also enjoy frozen bananas (as do my children). Freeze berries with natural yogurt in ice-cube containers for a fantastic summer treat. Also hang a full yogurt pot, with a hole cut in the bottom, so the yogurt drips out. Natural yogurt is very good for the chickens' digestive system.

Do not feed treats that are high in sugar or salt, and always cook potato skins. Raw or undercooked beans and pulses are poisonous, as are avocados and rhubarb leaves, which contain a fungicidal toxin called persin that has been known to cause cardiac distress and heart failure. Raw egg should never be given as it can encourage the girls to egg eating. All chickens have different likes and dislikes, so see what works for yours.

TOP LEFT: Home-grown corn on the cob provides great entertainment when hung up. TOP RIGHT: Apples hung in a bird feeder. BOTTOM LEFT: Hooped cereal threaded on string provides hours of fun. BOTTOM RIGHT: Sunflower seeds are a delicious treat.

Choosing your chickens and getting them home

Choosing your chickens

TOP LEFT: The Rhode Island hybrid is a popular choice to start with. **TOP RIGHT:** A hybrid utility Light Sussex is a fanastic layer. **BOTTOM LEFT:** Hybrid White Leghorns lay pure white eggs. **BOTTOM RIGHT:** The hybrid Black Rock is a stunning bird.

I have kept many other breeds but the ones I have mentioned here hold a special place in my heart. Remember to do your research and speak to other poultry keepers before deciding on what's right for you. We all have different likes and dislikes, so what's right for me might not be best for you.

Utility breeds are reared for eggs and, when they become less productive, you can eat them. Orpington, Brahma, Rhode Island Red, Plymouth Rock, Jersey Giant, Indian Game, Faverolles and Houdan are just a few examples of pure breed utilities. I have eaten some of the cockerels I have reared that I didn't need to keep. I tend to use commercial meat birds which you can buy very reasonably from commercial rearers. They grow fast and are hardy, and taste fantastic, making them worthy of a place in the garden. If you wish to rear your birds for meat, the best tip I can give you is DO NOT NAME THEM! I am a real animal lover but I am not vegetarian. I have a lot of comments from people who are horrified to find that I eat my own chickens. All I can say is that my girls and boys have a lovely free-range life, are treated with respect and are thoroughly enjoyed. I can also hold my head up and say that I have no part in battery eggs or intensively reared meat. The first time you sit down to dinner to eat one of your own birds is hard, but the immense satisfaction of a job well done soon helps you to forget the not so nice part.

The best age to get your chickens is at point of lay – from 16–21 weeks – especially if you're a complete beginner. You can be certain that they're hens and not cockerels and no special care or diet is needed. They usually start laying, depending on the season, at around 19–21 weeks. They are also still young enough to tame.

HYBRIDS

Hybrids are available worldwide and are chickens that have been bred mixing different breeds together to produce one that excels at egg laying. They are very hardy, healthy birds. All brown egg layers are based on the Rhode Island Red. All white layers on the White Leghorn.

• **Black Rock:** Stunning birds with a lovely iridescent sheen, these are happy as long as they are well fed.

- **Blacktails**: These are very friendly and great egg layers. They have a black tip to their tails and light feathers.

- **Blue:** A plump chicken that walks around with a regal air, and usually has a lovely fluffy bottom.

- **Goldline:** A traditional-looking farmyard chicken, she is cheeky and enjoys human interaction, regularly breaking into a trot to catch your attention.

- **Light Sussex:** Not to be confused with a pure-breed Light Sussex, her neck and tail markings are much bolder. She is a lovely girl, no trouble and a fantastic layer.

- **Magpie:** A truly stunning gothic-looking chicken that is sometimes referred to as a Sussex Reverse. Lovely nature.

- **Rhode Island hybrids** are a popular choice to start with. They are very good layers – expect up to 300 eggs a year given the right food and housing conditions – and come in a variety of colours, from black to ginger, speckled blue and off-white. They all make lovely placid family pets. They will give you different coloured eggs, ranging from pure white to dark brown and even blue. Hybrids from different suppliers are given different names. Below is a description of their personalities from my own observations. Take into consideration that all chickens are individuals, and there is always one that proves me wrong!

- **Silver Link:** Cream in colour, sometimes with ginger freckles, very friendly and quite attention-seeking ways. Certainly not bashful.

- **Skyline**: She's a busy chicken with a need to be top girl. Not the friendliest of birds but lays stunning blue eggs

- **Speckled:** A docile and gentle chicken that looks very similar to pure-breed Marans, and sometimes has a few feathers on her legs. A real lady..

- **White Leghorn:** A beautiful pure-white chicken with a larger-than-average comb which flops over to one side, rather like a bonnet. Rather nervous and better in a free-range environment. Reminds me very much of a slender ballerina. Lays pure white eggs.

ABOVE: A hybrid Speckled hen.
RIGHT: A chamois Poland cockerel.

PURE BREEDS

Pure breeds come in an abundance of shapes, sizes, colours and personalities. They are classed as pure breeds because they breed true, meaning their young will always resemble their parents. All pure breeds have their own standards (what the breed is required to look like, i.e. the shape, colour and size) which in the UK are set out by the Breed Club, which is affiliated to The Poultry Club of Great Britain (PCGB). Many are available worldwide. The following are pure breeds I have kept myself:

• **Appenzeller** is the national breed of Switzerland. These chickens are real free-rangers, preferring the open land and roosting in the trees than the confinement of a coop, and are very good fliers. They are rather highly strung, so are not the best choice for a small garden if you have children. If you want a lot of interaction with your chickens, these are not for you. They lay a reasonable amount of smallish white eggs.

• **Araucanas** are renowned for laying blue eggs, but in fact they range in colour from blue to green to pink! They are fun and have an upright gait and funny-looking ear tufts. They come in lots of colours, lavender being the most common. They adapt well to being confined and lay a very good amount of eggs. Having their eggs in your basket certainly causes a stir!

• **Barnevelders** lay lovely dark eggs, and lots of them. They are somewhat lazy so you need to watch their diet, as they can become fat. Being large, they can be kept with a low fence (but remember it won't keep out predators). They are well suited to confinement and are remarkably hardy. I had a stunning group of double-laced girls which were very easy to keep.

• **Brahmas** are my absolute favourite of the heavy breeds. They have feathery feet, amazing bottoms and an expression that always makes me giggle. They are very calm and placid birds and make wonderful pets. Their egg size is rather disappointing considering their size but they lay around 140 eggs a year. Brahmas do go broody and make lovely mums, but they have a tendency to tread on their eggs and break them, and can accidentally do this to their young too. Brahmas come in an array of colours, my favourite being gold partridge. My trio of Brahmas have made many people gasp at their beauty. Brigadeir, my particularly gorgeous cockerel, was a real head-turner as he marched through the yard, and, being as large as a small child, he had a certain presence about him.

PURE BREEDS

Hard feathered breeds are game birds with tight, short feathers that follow the shape of the body. They include: Old English Game, Malay, Aseel and Modern Game.

Heavy breeds have lovely temperaments. They were developed for both eggs and meat and tend to eat a lot. They include: Brahma, Orpingtons, Dorking, Cochin and Faverolles.

Light breeds are good egg layers, but not table birds. They can be flighty and nervous. They include: Araucana, Ancona, Leghorn and Appenzeller.

True Bantams are naturally small birds and are usually only kept to show or as pets. They are great if space is an issue. Examples are: Japanese, Pekin, Sebright and Belgian.

Rare breeds have no club and are in decline. They include: Andalusian, Norfolk Grey, Houdan, Orloff and Vorwerk.

LEFT: A Red Millefleur Pekin hen looking very proud.

• **Buff Orpingtons** are stunning and are often featured in magazines. They have lovely personalities and are calm and inquisitive. Because of their size they are not flighty nor escape artists, but they need larger houses with low perches. Their feathers are gorgeous and soft. They come in black, blue, buff and white. They do go broody and make lovely mums. For a pure breed their eggs are large and they produce around 160 in a season, but because of their placid nature they can be bullied by other breeds.

• **Campines** are beautiful birds and also quite wild. They are not a breed to keep if you are short of space. They have a very unsettled personality and even with patience they remain wild at heart – believe me, I have tried every trick and bribe I know. They do not make good broodies but lay an abundance of white eggs throughout the year.

• **Cochins** are very large with feathery feet and a lovely bottom! When you watch them pecking around it looks as if they are wearing bloomers. These chickens are friendly, contented and easy to keep. They need a mud-free run due to their feathery feet. The eggs they lay are small considering the size of the girls, and they lay about 115 eggs a year. They make lovely mums. They come coloured black, blue, buff, white, partridge and mottled. Rocky was my star cockerel who produced some fantastic offspring. He was a real gent until the spring came, when he turned into a nightmare, running at whoever entered his pen, launching himself at you spurs first. Because of his size and weight he could certainly pack a punch. These are dual-purpose birds, meaning they can be eaten as well as lay eggs. Their feathers are fantastic for stuffing cushions.

• **Faverolles** are very distinctive with a lovely beard and muff, and they have five toes. They are fast-growing with a friendly, tranquil nature and they lay around 115 eggs a year. It is not unknown for my girls to lay into the winter. Because they are so friendly and polite, they can be bullied by more boisterous breeds. These are my favourite to watch as they are so serene and sedate, and you can almost feel the stress of the day slip away. You can eat them and their meat has a lovely texture.

• **Leghorns:** I love these girls, they look like little ballerinas, and are very agile and always on the go. They lay lots of large white eggs and come in many colours. They love to range freely and are quite capable of roosting in the trees. Their combs and wattles are very large and sometimes need protection in very cold weather. The combs give the appearance of a lopsided bonnet balancing on a very petite head.

• Marans are prolific egg layers and they lay beautiful chocolate-brown eggs. They are more flighty than the other heavy breeds mentioned, but are friendly, busy girls that can be easily tamed. They come in a variety of colours, including cuckoo, black and copper/black. They make good mums and are very intelligent and alert.

• Polands are lovely and have a huge ball of feathers on their heads (crests). They are very comical to watch, extremely good pets, and a reasonable layer of small white eggs. They do need extra care because of their head feathers; they can be prone to eye infections and are best kept with their own kind. Their crest also impairs their vision, which makes them slow to react, so they are best kept in runs.

• Welsummers are very traditional farmyard chickens that lay lovely dark brown eggs, and lots of them. They are hardy and spend a lot of time foraging. They don't go broody too often, but when they do they make fabulous mums. I had several Welsummers a few years ago and they often used to trot into my sunroom to share my custard creams!

TRUE BANTAMS

Are the smallest chickens, with no large counterparts, and are very popular when space is an issue. Examples are:

• Japanese are really elegant chickens with very upright tails and slightly drooping wings. Their legs are very short and can hardly be seen. They are not very good in the cold or in wet weather as their chests can get very wet due to their tiny legs, so they are better suited to a sheltered area or being indoors. They make great, friendly pets so are extremely well suited to small children. However, they are not very good egg layers at all.

• Pekins are my favourite small chickens. They walk like Charlie Chaplin and have personalities that would suit a lion. They do need a dry run as their short, feathery feet can get very wet and muddy. They are hardy and robust, and mix well with most breeds if given enough space. They are good broodies and make very protective mums. Expect around 120 small-sized eggs a year, although this number really does depend on how much time they spend being broody. You will see Pekins in a vast array of colours.

ABOVE: A pair of pure breed Japanese Bantam hens. OPPOSITE PAGE TOP LEFT: A pure breed Black Silkie hen makes an attractive sight in any garden. TOP RIGHT: A pure breed Partridge Pekin hen. BOTTOM LEFT: A pure breed Lemon Millefleur Booted Bantam hen. BOTTOM RIGHT: A pure breed White Silkie hen is a characterful bird.

- **Sebrights** are compact chickens with fantastic tails and the lacing on them is truly stunning, almost as if they have been painted. To see one close up is amazing. Their characters are very likable and calm, but they do not lay many eggs.

- **Serama** is the smallest breed of poultry in the world. They are adorable but very vulnerable and they need protection from above. It is not unheard of for magpies, crows or hawks to take them. They have a fantastic personality and with their proud chests and upright tails, they remind me of little Victorian ladies! They lay very few eggs, which come in all colours. Despite their size don't be misled; the cockerels can make a real din in the morning.

- **Silkies:** I would never be without this breed in my garden. They are adorably fluffy and cute, and such endearing characters. Gertrude, a truly adorable girl (if not a little headstrong), used to be a regular in my kitchen. Her mannerisms even made my husband smile as she demanded treats and argued with our Jack Russell dog over the water bowl. Gertrude was often spotted on the back of a patio chair surveying her land. She raised many broods in her time – even ducks.

  Their feathers are fluffy, more like fur, and because of this their feathers can get wet very quickly. They also have blue skin and flesh, and a fifth toe. In my experience I have found Silkies to be very hardy and easy to keep. They do, however, spend a great deal of time being broody, so do not expect many eggs, which come in many different colours.

The tyrannosaurus rex is a very distant relative of the chicken.

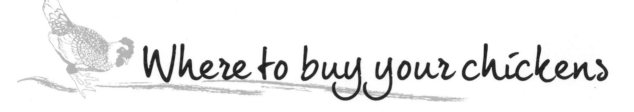

Where to buy your chickens

If you are a beginner it is advisable to go to a reputable breeder to buy your chickens. You should look around the establishment and be able to view your chickens and their environment. They should offer you advice and answer your questions freely. In my view they should also question you if they care about their birds. I can spend up to an hour with beginners going through questions and reassuring not only myself but my customers that the chickens are right for them. Most good breeders will give help over the phone. I encourage people to update me on their chickens. Never buy chickens that you have not seen.

A lady who had purchased some hens from another breeder came to visit me. They had become ill very quickly and when she phoned the breeder he wanted nothing to do with them. It transpired that she had not seen the birds or their environment but had just accepted the box of chickens, which were very underweight, lethargic and covered in lice. With lots of tender loving care and some antibiotics, wormer and lice powder they made a full recovery – but she has no idea what age they are or if they have been vaccinated.

Another person who had originally bought his chickens from me phoned to say that they were all poorly. The symptoms included a hunched appearance, being lethargic, sneezing and not laying any eggs. After a while I discovered that he had bought two hens that didn't look quite right at a market but he put it down to stress. He had put these in with my hens and they all became ill with mycoplasma (see page 123).

Beware, there are lots of people out there jumping on the chicken-selling bandwagon to make easy money. Make sure you stick with established breeders and go by word of mouth. Always check the chickens and their environment, and if you don't like the look of either, trust your instinct and walk away. I have been keeping chickens on a large scale for five years. It isn't easy and it doesn't make you rich but it is a way of life and a passion. I like to look after my girls as I would look after my children, and when you have up to a thousand it is a huge job.

DOS AND DON'TS

Do
- Handle and check the bird yourself
- Ask questions
- Look around
- Take your time
- Keep a minimum of three birds

Don't
- Buy a bird you have not seen and checked
- Buy one that you feel sorry for
- Buy if you are not happy with the way the chicken is being kept

RIGHT: A fabulous White Silkie cockerel.

TOP LEFT: Holding a chicken with her wings in. TOP RIGHT: A sure indication this girl has been pecked (note the V-shaped feathers left). BOTTOM LEFT: Comb of a Speckled Sussex. BOTTOM RIGHT: An inquisitive hybrid pullet.

WHAT TO LOOK FOR

Once you have chosen where you are going to get your chickens, here's what to look for when you check them. Don't feel silly, as a good breeder will be only too happy for you to do this. I do it with all my customers as a matter of course.

The chicken you choose should be quick and alert. Never pick a hunched-up, sleepy-looking one because you think she is docile! It usually means that she is unwell. It's normal for a chicken to run away when you are trying to catch her – this is not a sign of her being unfriendly.

Once caught, she may make a fuss for a while (I would if I was caught and picked up!), but this is perfectly normal. Talk to her and hold her comfortably. If you're not confident just ask the breeder to show you how. Then look at her eyes – they should be bright and clear, and should never have any discharge. Check her comb – if she is a young pullet (sixteen to eighteen weeks old) it will be small and pink; if she is slightly older it will be upright and red. Some breeds, such as the Blues, have a bluish tinge to their skin, and this is perfectly normal. Her beak should be a natural shape. If it has been de-beaked (where part of the beak is clipped when they are young to stop them pecking each other) it means she has been kept in cramped conditions and usually not allowed out to roam. Think: do you really want to encourage this practice? If people did not buy de-beaked birds, the rearers would be forced to think again. There shouldn't be any discharge or smell coming from the nostrils. The legs should be smooth and clean. The feathers need to be clean and glossy, with no bald patches, and the bottom (vent) should also be clean.

You can tell if a chicken is in lay by measuring the width between her pelvic bones. In a mature hen in full lay there should be a gap the width of about three fingers.

EX-BATTERY HENS

You can now purchase battery or spent hens (those at the end of their commercial life, which is usually one to two years) very cheaply. They will be lacking in feathers, not know how to perch and be very nervous of the great outdoors. But in a short time and with plenty of love, patience and good feed, they will re-feather and gain confidence. People find this very satisfying, but it can be emotionally draining and stressful if you are new to chicken keeping.

You can call a chicken a chicken regardless of its age or sex.

Getting them home

RIGHT: A traditional
movable chicken coop.

Before you go to collect your chickens, it is very important that their coop and run are ready for them. Ensure the coop has sufficient bedding in it and that there is feed – preferably the one the chickens are used to – and cold water. If you haven't visited the place you are going to get your chickens from (which I advise that you do), phone them beforehand to ask if they could let you have some feed to bring home with you. The chickens do not need the stress of moving and new food. Peel and slightly crush two cloves of garlic and place them in their water – this is great for a chicken's immunity, which will be lowered with the stress of moving. Collecting your chickens is always very exciting but explain to any children coming with you the need to be quiet on the journey home.

One memory I have is of a lovely family with three excitable children coming to collect their chickens from me. All was going well until the father placed the box containing the three chickens on the knee of his eldest daughter in the car. This was greeted with hysterical screaming, and whilst arms were flailing I managed to take the box out of the car. While the parents calmed the situation, I firmly explained how the noise and behaviour was unacceptable as the chickens would have been truly traumatised. Whilst I can fully understand the excitement of the children, the first step of good chicken keeping is to maintain a quiet and calm manner, and this needs to be explained from day one. Respecting any animal is a must.

TOP: A lovely alert hybrid hen.
BOTTOM LEFT: An ideal box
to transport your girls home in.
BOTTOM RIGHT: A point-of-lay hybrid
ready to go to a new home.

TRANSPORTING YOUR CHICKENS

Pet carriers are very good for bringing your chickens home in, but if you haven't got one a cardboard box is also excellent. It needs plenty of air holes cut in it so the chickens can breathe – these can be made simply by pushing the scissors in and turning. Shredded newspaper makes good bedding and can be disposed of when you have finished with it. It is best for your chickens to travel together in one box if possible, as it is less scary when they're with their friends! It is not acceptable to tie the chickens' legs together and put them in the boot of the car, neither is placing them in brown sacks. One family thought it was fine to let their children hold them on their laps in the car, although boxes were provided. This of course would be dangerous.

SETTLING THEM IN

Once home with your girls place them into the coop and leave them quietly for about an hour, or half an hour if it is a hot day. Be careful when opening the box to transfer them to the coop, as they can sometimes jump out. I can remember one Saturday taking a telephone call from a lady who had bought her chickens from us that morning. Her White Leghorn had jumped out of the box and run off. She thought it might run back to me, but I explained it was highly unlikely as she lived 20 miles away!

Keep other pets and children away, and after about an hour you can let the chickens out. They might well be very reluctant at first, but resist the urge to push them. Be patient and they will eventually come, especially for a treat which can be popped in if they are being particularly stubborn! If possible sit with them for a while so they can get familiar with you.

Once they have been out and fed and scratched around, calmly introduce any pets, but do it responsibly and carefully. It can be a gradual process and occasionally takes a while, but if done calmly and confidently all should be well. Screaming and overreacting to a situation will just cause more stress and anxiety.

THE FIRST NIGHT

Sometimes chickens can be quite confused about going to bed on their first night. If you place them in their coop and leave them for a while to come out of their own accord, they will usually remember where to go. Some girls can be very nervous and have no idea. The only option is to physically pick them up and pop them in yourself. You should only need to do this at most for a couple of nights. Another option is to place a torch in the coop so the light can be seen through the pop hole. It does work but again not on all girls.

Handling

Handling chickens is a very important part of poultry keeping and should be done regularly and with confidence. When you purchase your chickens, ask the seller to show you how to do it. One of the best ways to learn the art of handling correctly is to watch it being done. Hopefully this will make everything clear and you will go away happy with your new knowledge. The key to handling your chickens successfully is to do it slowly and confidently, while talking to them softly. Any lunging, shouting or fast movements will alarm them and should be avoided. Handling them at night is always preferable, as they will be calm and less likely to fly off. Talk to them as you approach and handle them gently. You are aiming to slide your hand in under their breast so you end up cradling them. Place your thumb over one wing, and tuck the other into your body. The more practice you have of this, the better you will become. If it is done calmly your birds will soon realise you mean no harm and react accordingly.

Chickens are very sensitive to their immediate surroundings. They can become stressed if kept in a coop and run of inadequate size, if they are handled roughly or if they are living within close proximity of young or unruly dogs. Do you have young children that tease them or are loud? Lack of food and water will cause them stress, as will very hot weather with no shade, or extremely cold weather with no shelter. Is Mr Fox pestering them at night scratching at their coop? This, too, will make them anxious.

I once had a phone call from a lady who was very perturbed because her hens had stopped laying for no discernable reason and she had only had them two months. It transpired that her Labrador had been chasing them and had actually caught one. I explained that although none were hurt, the stress of the incident would certainly have caused the lack of eggs. Chickens can become used to dogs barking and children playing, even the noise of helicopters overhead, but anything that makes them uneasy can cause them to become stressed and stress can affect their health. They will remember stressful incidents for a while, but it is perfectly possible to rekindle the lost trust with time and patience.

RIGHT: How to hold a chicken correctly, making it feel safe and secure.

CHILDREN AND CHICKENS

Having grown up with chickens, my children are all rather blasé about the whole thing. I found it interesting when we had friends round and one of their children asked why my chickens' eggs didn't come with the date stamped on!

Children are normally far more resilient than we give them credit for and keeping chickens has given me the opportunity to deal with subjects such as death, illness and reproduction. We have had tears and laughter.

My youngest learned a very hard lesson on gentle handling when his excitement got the better of him and he accidentally hugged a chick to death – something he will never forget or do again. But death is something we all have to deal with at some point, and encountering it with a pet allows children to understand that it is perfectly natural.

We have had all sorts of pets in our house, and I can honestly say that chickens are the best. They let you cuddle them, they follow you, and even join in when you are in the garden, in their own way. And they produce eggs for breakfast.

The children have also benefitted at school, as a lot of schools are now showing children about incubation. My youngest daughter's confidence was given a huge boost when she became the chicken guru in her class. I have had several visits from local schools and the joy and enthusiasm shown by the children gives me a real buzz. A little boy who was buying some hens for himself asked me why farmers think it's alright to keep chickens in cages. Both his mother and I were amazed that a small child realised this is wrong. A couple of boys who bought chickens from me are making a reasonable return on their investment by selling the eggs. They even phoned me to ask if they could buy trays of eggs from me because they couldn't keep up with the demand and were getting a better price than I was selling my eggs for. I definitely see them as entrepreneurs of the future!

Children are perfectly capable of looking after chickens but they do need a responsible adult overseeing them, and occasional nagging to do the chores.

OPPOSITE AND ABOVE: Children love chickens and benefit from the responsibility of owning a productive pet.

There are more chickens on earth than there are people.

Introducing further chickens to your flock

The important point to make is that unless they are from the same place that you obtained your original chickens from, you should keep them separate for up to a month to observe them and make sure they are disease free. When you are sure that they are healthy, it's time to introduce them to the existing flock. If you have a lot of space or your girls are free ranging, I would opt for doing this at night when the older girls are settled. Sprinkling red mite powder or garlic on them so they all smell the same will help. In the morning when you let them out, put extra food and water dishes temporarily in the run to dilute any territory issues, hang up some distractions (see page 34), add a few logs and change the run as much as possible, so the older chickens are slightly unsure as well. Make sure you're around for the day, to keep an eye on them. If you have a cockerel in your flock, the hens will normally settle quickly. A flock consisting of all girls can take a good week to settle. Introducing new chickens to a small run is slightly more stressful, both for you and them. Distractions and treats help to occupy the girls but will not eliminate the arguments as they try to establish the new pecking order (see page 94). If possible, make a second small temporary run near the main run so the girls can see each other and sleep together for a while before the final introduction. I find moving the new chickens into the original run, and the old ones into the temporary run for a day or so will also help. It is always best to introduce two chickens at the same time for obvious reasons.

Introducing chickens to other pets

When introducing your dog, have it on a lead and walk around the enclosure. If it barks and lunges tell it off firmly and walk away from the coop. It does take a while for dogs to settle with something new but normally one peck from a chicken is enough. I have a Jack Russell and an English Bull Terrier, both of which wouldn't dream of touching any of my girls. I am, however, a great believer of firm handling with my dogs. My Jack Russell has even been known to share her dinner with Gertie, one of my Silkies. Remember it is never advisable to leave your dogs totally unsupervised.

Cats are normally fine with chickens – if you log on to YouTube, you will see that it is the chicken that always appears to be in charge. There are, however, instances when a local tom might cause problems. It is not a common problem but one to bear in mind. Can you keep rabbits and guinea pigs with chickens? Yes, but you need sufficient space and areas where the guinea pigs and rabbits can be by themselves, and the chickens cannot get to them. A peck from a chicken can hurt a small animal.

RIGHT: My Jack Russell enjoying an afternoon snack with the girls.

Raising your own chicks

By far the best way to hatch and rear chicks is nature's way – a broody hen. Silkies, Pekins, Sussex and certain crossbreeds are fantastic mothers and are a joy to watch. The other option is to use an incubator. There are some very good ones on the market, so do your research and buy the best one you can afford. Bear in mind that eggs need turning at least three times a day at regular intervals. An automatic incubator will turn the eggs for you. Read the instruction to your new incubator thoroughly and follow them. You will need to place it in a room where the temperature is constant, away from direct sunlight which will cause the temperature to rise. Disinfect your incubator before use and run it for at least 24 hours, checking the temperature and humidity levels before use. The humidity level needs to be around 53 per cent. Once the correct levels have been reached, you can place your eggs inside. Eggs should be as fresh as possible and disinfected with a product that is suitable for both eggs and incubator.

Candling

The temperature should be 37.5°C for the first 17 days. The humidity needs to be around 53°C up until day 17. Those levels are from my own experience using my incubators. Always read your manufacturer's guidelines and follow these, as the levels will vary according to the type of incubator.

When the eggs have been in the incubator for about six days you need to 'candle' them. This involves holding the egg against a bright light so you can view the contents. You can make your own by placing a torch inside a cardboard box that has a hole cut in it. It's best to do this in a darkened room. It is important to check the eggs' fertility at this stage, as infertile eggs can explode and contaminate the incubator.

Candling also allows you to monitor the humidity levels by checking the air cell in the egg. If the air cell is too small it means that the humidity is too high but if it is too large the humidity is too low, and you will need to adjust the level accordingly by adding or taking water away. Eggs that are not fertile will be clear. Those that are fertile will have a small, dark blob with veins coming out, rather like a thin petal on a flower. Eggs that have a dark circle inside can also be discarded. I candle again at 14 days to check that everything is alright, and then I leave the eggs alone.

RIGHT: Gorgeous fluffy chicks; they are a joy to raise but remember that at least half will be cockerels.

ABOVE: A mixture of pure breed chicks chilling out on my daughter's lap.

Hatching

At 18 days, stop turning the eggs so the chicks can get into the correct position to hatch. The first sign of hatching is called 'pipping' – this is where the chick breaks through the membrane and makes a small hole in the shell. You will be able to hear the chicks chirping to each other, and if you speak to them they will answer you!

It can take a chick about 24 hours to hatch from it first pipping to coming completely out of the shell, though some seem to do it in a matter of hours.

It is not advisable to help out the chicks – there is usually a reason for them having difficulty. It could be that there are problems with the temperature or humidity levels or simply that it was not meant to be. Late hatches (when the chick takes longer than 21 days) are usually because the temperature is too low. It may not grow well, have thin legs with crooked toes, or might not be able to learn to eat or drink. If the temperature is too high the chicks will hatch early (before 21 days) and they can suffer from splayed legs (see page 132). If this happens you will need to adjust the temperature and humidity levels when you hatch again. It is good practice to keep records of any changes you make and what the results are. This will mean you will eventually reach your optimum formula to get your best hatch. It will take time. I have had my incubators for years and know them as well as my husband.

If there is a power cut, eggs can withstand drops in temperature better than increases, so don't panic. If the power is going to be off for a short while, cover the incubators with towels or an old blanket/sleeping bag to prevent heat loss. Ring your supplier to check how long the power is going to be off for. If it is for a long while, place the eggs in a box, wrap it in a towel and put it in an airing cupboard, by the fire (not too close), or near the Aga (if you have one). Your hatch rate might not be very good but it's worth a go, especially if the eggs are ready to hatch.

I have had a chick hatch in my bra before. When checking on one of my broody hens one morning, she had given up and left her nest. Two chicks were half out of the shells but stone cold and lifeless, another one had a faint chirping coming from it, so I popped them all in my bra and continued to go about my morning jobs. By about ten o'clock the chirps were becoming louder and more determined. By two in the afternoon I had two lovely fluffy girls scratching around, closely followed by number three a couple of hours later. Needless to say, they didn't stay there for long!

It is truly amazing to watch a chick hatch – even my teenage son cannot resist a look and is often found in the incubator room, saying he's looking for a bicycle pump.

Rearing

Once your chicks have hatched and dried in the incubator, they should be moved into their new home. I find rearing them in large plastic boxes works very well for me, as they are both easy to clean and draught-proof. I line the bottom with a layer of newspaper and for the first week I place kitchen roll on top. This prevents the chicks developing splayed legs due to the floor being too slippery (see page 132).

It is important to keep the chicks warm. This is done by suspending a heat lamp above the box. You will know when your chicks are at the right temperature by their behaviour. If they are all huddled together under the heat lamp they are too cold; if they are all round the edge of the box, trying to get away from the heat, they are too hot. Both scenarios need to be avoided.

Chicks kept at the right temperature will be happy walking around and sleeping where they drop. Most heat lamps come with a chain to hang the lamp from, which makes adjusting the heat very easy. Simply raise the lamp if the chicks appear to be too hot or lower it if they seem too cold. So just watch your chicks and adjust as necessary. You are aiming to harden them by about four to five weeks, but this will depend a bit on the weather and the time of year. You harden your chicks by gradually reducing the heat, which is done by raising the heat lamp by about 2.5cm a day. Just observe them and they will let you know if they're not ready.

Ideally hatching is best done in the spring, as Mother Nature intended, because of all the natural benefits, good weather, fresh grass, an abundance of bugs and longer days to eat and grow.

On the fourth or fifth week, turn off the lamp during the day, but pop it back on at night for a week. Then stop the heat lamp altogether. If you are rearing your chicks in the colder months, you will have to use your own judgement on lengths of time, but remember the tougher they are from an early age, the hardier the adult bird.

Chicks need to be fed on chick crumb, as this is very small and it contains all the nutrients they need. Check the label and make sure it contains anti-coccidiosis. Coccidiosis is a disease that can be fatal in chicks (see page 128). I place the crumbs in chick feeders, but for the first couple of days I also sprinkle a little around the chicks and try to encourage them to eat, rather like a mother hen would do. They don't take long to get to grips with eating and drinking.

When they are six days old, stop using kitchen roll and pop in a small amount of sawdust instead. I use horse bedding – the sawdust is much larger and the chicks tend not to eat it, which can happen when they are kept on very fine sawdust. Clean them out regularly, as dirty, damp litter can cause multiple problems.

PROS AND CONS OF REARING CHICKS

Pros
- Great fun and rewarding
- Educational for children
- Produces great chickens

Cons
- You get cockerels
- Time consuming
- Requires additional housing
- Deaths due to disease and/or power cuts

Do not count your chickens before they are hatched
- Aesop

You need to make sure that water is available at all times. Special chick drinkers are available that work very well. Otherwise you can use small plant saucers, but make sure you add clean pebbles to prevent the chicks from drowning. Drinkers should be placed on raised wire. This can easily be done – screw four pieces of batten together to form a small square and tack wire tightly over the top. This stops the sawdust clogging them up as the chicks scratch.

Change the water regularly – about three times a day – as it can become warm under the heat lamps. Add very small amounts of greens. I tend to favour grass and lettuce, cut into very small pieces and sprinkled on the floor. This also helps to avoid pasting (when droppings stick to the chick's bottom; see page 132). After about a week, start adding very small amounts of flint grit to their food to help the digestion process. Don't add oyster shell as the extra calcium in the diet at this age will cause bone development problems, and can also damage their kidneys.

At six weeks the chicks should be out in their own coop. If possible move the run regularly to avoid any problems, and protect them from the elements as you would adult girls.

They should now be fed with grower's pellets until they are 16 weeks old. Do this by gradually mixing half chick crumbs and half grower's pellets for a week. Add a peeled clove of garlic to the water in the container. Poultry tonic can also be beneficial, especially if the weather takes a turn for the worse. Worming is a must when you put the chicks into their outside coop. Verm-X is my preferred wormer, and for chicks I use the liquid form, which can be added to the drinking water or soaked into bread. Make sure the chicks are in a covered run, as they can easily be taken by aerial predators such as sparrow hawks, crows and magpies.

Daily
care and
behaviour

Routine

TOP LEFT: A lovely Pekin hen.
TOP RIGHT: Topping up feeders with
a scoop makes life much easier.
BOTTOM LEFT: Collecting eggs daily
reduces the chance of accidental
breakages. BOTTOM RIGHT: Different
coloured eggs make an attractive
display in any kitchen.

So you've set up the coop and your chickens have been happily introduced to their new home. Once settled, you will find that your girls will thrive on routine; it makes them feel safe and confident so start the way you mean to go on. Your daily routine will continue throughout the year. The only change will be the times at which you do things, which are dependent on the season. There is no need to set your alarm and be up at the crack of dawn. If anything, I advise you to wait until the world is awake in the hope that Mr Fox has gone to bed, but bear in mind that they're not teenagers, so they won't appreciate being left in until midday. The time they go to bed also changes according to the seasons, so I always go out first before dark and watch and wait.

It won't always be possible to stick rigidly to your routine and the occasional deviation is fine – if you are planning to go out for the evening, for example, put them to bed before you go, they won't be that keen but much better this than being eaten by the fox.

Once you've unlocked your chickens they will need fresh water. They drink a lot, so in hot weather you may need to replenish their supply in the afternoon as well as in the morning. I believe in feeding my girls on a regular basis; I top up their feeders every day so they have food available all the time. The girls eat what they need and in my experience are not overly greedy. Their food intake varies according to weather conditions, with them requiring more food in the winter to stay warm. Layer's pellets have everything in them to keep the girls healthy and produce delicious eggs. With this done just cast an eye around the run perimeters to check that no predators have been around.

After your chickens have had their breakfast and a stretch they will go off to lay their eggs, so give them an hour or so and then collect them. You should avoid letting the eggs build up over a few days because if you do they can break and, once broken, the chickens won't be able to resist eating them. Once started, this habit can be very difficult to stop.

In the afternoon/early evening your chickens will appreciate a treat such as a handful of mixed corn. This is a great time to tame them. Just call them over or shake the bucket before scattering and they will associate the noise with the treat and come running. If your chickens are out and you need to get them back in, throwing a small handful of corn on to the floor of the run will tempt them back in and it makes them scratch and forage, a good, natural behaviour. It also gives them something to see them through the night.

Remember, though, that to chickens corn is like 'chocolate' and fat chickens do not lay many eggs! I recently had a customer come to the farm complaining that one of his girls wasn't laying. When I took the box from him to have a look at her I thought he had several hens in there, but instead there was just one very happy but enormous girl. After a strict diet she started laying again.

If you have any scraps of vegetables, fruit, pasta, rice, etc., mid-afternoon is the time to feed them to your chickens. If possible, hang them so the girls have to reach for them – a bit like a workout. This is a lovely time to watch their antics, establish who the boss is and check they are okay. Just before dusk, lock your girls up. In the early days of having them, you may find that they are slightly confused and need some persuading. Calmly and slowly ushering them with open arms usually works.

Once they're in bed, it's always a good idea to clear away any uneaten scraps and treats so they don't attract rodents. Double check the lock and the job's done.

This is much easier to do without a glass of wine, as my friend happily does every night to escape the chaos of the house!

If you are around in the garden during the day the girls will always appreciate a trot around the garden with you, but be warned they will eat your plants and once settled will think nothing of going into your house for a nose. You should never leave them in the garden out of their run without you being around as it only takes a minute for Mr Fox to pay a visit, as I know from personal experience.

TIMETABLE

Early morning: Unlock your chickens, feed them and give them fresh water.

Mid-morning: Collect eggs.

Afternoon/early evening: Offer your chickens a treat.

Dusk: Lock up your girls, clear away uneaten scraps and check the coop perimeter.

RIGHT: A hybrid French Maran enjoying a supervised afternoon free range.

GOING ON HOLIDAY

The first time you leave your girls to go on holiday is worrying but can be simple with the correct planning. If you are going away just for the weekend they should be fine if left with ample food and water. This is, of course, presuming the house and run are fox-proof. Thorough checks of the coop and perimeter should be made before you leave. It is a good idea to hang some vegetables in the run for their entertainment.

If you plan to leave them for a week or more you will need to give it slightly more thought. Neighbours are usually willing to check daily and replenish food and water in return for the delicious eggs. But if neighbours or friends can't help out, try putting an advertisement in the local post office or newsagent. Pet shops may know of people who will come to your house or provide lodging whilst you are away. Be cautious of the latter as, if other poultry are at the temporary home, diseases can be caught.

The day before you leave, give the coop a good clean, then you need not worry until you come back. Check that you have left enough food, grit and treats to last and that you have explained the routine to the person caring for them. The chickens may become bored as they will not be out roaming, so hanging treats and toys is a must to keep them amused.

It is also a good idea to leave a simple checklist for the chicken carer of dos and don'ts, plus the number of the local vet in case of problems.

RIGHT: An inquisitive hybrid Magpie.

Care and maintenance

WEEKLY MAINTENANCE

Once a week you need to give the coop a good clean. Empty out all the bedding and scrape any droppings off the perches. Have a good check for mites and if none are found you just need to air the coop for an hour or two. Make sure the pop hole is closed as the girls will not miss an opportunity to escape! On one particularly lovely day I left the door and nest boxes open in one of my large laying houses that had 50 girls in, and didn't close the pop hole. So I had 50 girls scattered everywhere. Luckily when I approached the chickens with a bucket of food they came running and followed me like the Pied Piper.

When the coop has had a good air, replace the perches and place fresh bedding inside. In the hot months I also sprinkle mite powder inside just in case. Remember to open the pop hole. Also disinfect your drinker – in hot weather it will become slimy and green. If you have a fixed run, rake over the ground and if possible add a log (see page 38) – something to make it slightly different for the chickens. If it's a movable run, place it on fresh ground – this is much easier to do when the girls are loaded into their house. Check your perimeters for any signs of wear and tear or predators. Top up the feed, and your job is nearly finished. It is advisable to regularly check where you store your bedding and chicken feed; this is often overlooked, until, one day, you are greeted with a gnawed storage bin and a pile of droppings. Move your containers and check for signs of droppings and chew marks, lift up the bedding and have a good look for any signs of rodents. If you find any you need to put down some bait, set a trap or call Harry (see page 87)! Put all your tools away, sit down with a cup of tea and watch your girls' excitement over their clean home. We live in a busy world, but I cannot emphasise enough the joy your girls will give you if you spend time with them. Give it a go, you won't be disappointed – it's much better than daytime TV – and while watching check that they all appear in good health.

If you have chosen a wooden house with a permanent floor, it's a good idea to purchase some cheap lino. Cut two pieces to size so they fit in the bottom of the coop. When cleaning you can roll the sides up and pop the mess straight into the compost or bin. The second piece can be placed into the coop while you clean and dry the dirty piece. Use a paint scraper to clean off the chicken droppings.

REGULAR TASKS

WEEKLY
Thorough clean, check for red mites, clean drinkers, check for signs of rodents/predators.

MONTHLY
Weather permitting, clean and disinfect, treat girls and house for red mite, disinfect drinkers, air the house thoroughly, worm the girls if necessary using Verm-X.

YEARLY
Treat the house with wood preserver, general maintenance of the run and house, hinges, locks, etc.

TOP LEFT: Some young Buff Orpington pullets checking out blackberries hung up for entertainment. BOTTOM LEFT: Verm-X natural wormer is easy to give and safe. FAR RIGHT: Home-grown lavender. My daughter dries it out and makes into lavender bags which she pins inside the coops to make them smell nice and help the chickens relax.

Monthly Maintenance

The first day of every month is a very busy one on the farm. It's deep-clean day, health-check day and worming day. This may sound daunting but it is great fun, and after years of doing the same routine, I still get a buzz and sense of achievement when all the jobs are done – and I certainly sleep well afterwards.

You need to move your house if it is movable, remembering to keep your chickens locked up in the coop. Disinfect your drinker and feeder, refill them and pop them back into the run. If you use Verm-X wormer, as I do, put the dose into the feeder. Once this is done, catch your girls one by one and give them a good check, looking for signs of mites and injury. Are their feathers looking good? Is their comb nice and red? Are their legs nice and smooth? How does their weight feel?

You will, after doing this regularly, begin to know what is right for your birds. Before popping them into the run, dust them with red mite powder and check to see if there are any problems.

The girls will now be happily having breakfast, so you need to empty the coop of bedding, and check for signs of red mite. If the weather is dry and sunny, spray the house with red mite liquid and allow it to dry by airing. When it is dry, scatter Diatom Powder around (see page 125) and place some fresh bedding inside. Open the pop hole and check all the hinges and locks – if they are starting to rust, smear them with grease. Check if the perimeters of the run are fixed and look for signs of rodents around the coop, particularly if it is on the ground. Deal with any problems as you find them. Do not be tempted to leave it for another day as you may lose your girls.

If you have a fixed run, you need to rake it over thoroughly, removing any heavily soiled areas, and spray with a disinfectant. If it's very wet you might need to add wood bark to make the environment nicer for the girls. While they like paddling in mud, they don't wipe their feet and will traipse mud into the coop which will add to your cleaning tasks. Wood bark is not expensive and works very well – the girls can still scratch and forage – and don't forget to add some enriching toys (see page 38). If you have a dust bath, top it up and add red mite powder to it. Finally, disinfect all the tools you have used, and put them away, checking your storage area for rodent activity.

Can I bath my chickens?

Yes, but usually chickens keep themselves clean and do not require any help. When people show their birds they usually bath them beforehand. If you are going to wash a chicken, have two bowls of warm water ready. Place the bird in one, wet it thoroughly and use a very small amount of baby shampoo to wash it. You need to do this calmly, talking softly to your bird. Try to hold the bird with one hand and wash it with the other – this takes practice and you will get wet if you do not keep the wings tucked in. There is definitely a knack to this. Rinse well and gently towel dry. You can use a hair dryer on a low setting following the direction of the feathers. Hard-feathered breeds are best left to dry naturally. Remember bathing is usually only done when showing the birds. If you bath those you keep as pets for eggs, the stress can stop them laying.

RIGHT: Blow-drying a Poland hen after her bath.

LEFT: Suspending a feeder in the run
helps to prevent the girls spilling food
and attracting vermin.

PEST CONTROL

It is said that we are never far from a rat. Rats and mice can be a problem when keeping chickens, but only if you are messy with their food and do not clear away uneaten treats. It is this that attracts them, not the chickens. Rubbish left around the coop and log piles can provide perfect nesting sites, so keep these areas clear, and put stored feed in metal bins to avoid any trouble. You are legally required to keep rodents under control and because they breed so quickly this can be a constant battle. Methods of control are: traps – either live-catch ones, which catch them humanely, or impact, which kill – bait, which poisons them, or Harry, my pest control man!

Traps can be purchased from most hardware stores, pet shops and DIY stores or via the internet. You need to follow the instructions carefully. If you are using live-catch traps, think how you are going to kill the pest once you've caught it. You need to kill them humanely – either by shooting them or a single blow to the head (which is much easier said then done). Drowning them is not allowed. Impact traps work very effectively but you need to be careful when setting them as you don't want to catch birds/cats/dogs/your child's guinea pig.

Traps need to be placed where you can see that the rats have been active – they are very wary of new things, so rub your new traps in your chickens' bedding, the compost heap or leave it outside for a few days. Do not site rat traps in the open as they tend to make their runs in sheltered areas.

There are lots of poisons on the market to use as bait and I have tried most. They work very quickly (between one and four days) but they present a risk to other creatures. Also you cannot predict where the rat will die – it could be under the floorboards of your house. Poisons need to be used very carefully. You must follow the instructions on handling and usage to the letter.

Official UK guidelines on the disposal of the rodent carcasses state that you need to wrap the body in a plastic bag, secure it, then place it in a rigid container that can then be put in the refuse. Elsewhere, regulations may be different, so check. Always wear gloves, as rodents can pose a health risk even when dead.

For a very reasonable fee Harry has taken on my vermin control. His knowledge and equipment is a godsend and I wish I had found him years ago. Bear in mind that by the time you have bought your traps or bait, and spent time positioning them, waiting and maybe catching a bird or two by mistake, that it might be more economical to find a local pest control man.

Over the years I have used all the methods of rodent control mentioned, even paid my sons money per rat shot dead. They used to sit silently for hours with their airguns earning their pocket money. I have had great success with all the methods mentioned.

ABOVE: Placing concrete slabs around runs helps prevent predators digging their way in

PREDATORS

Mr Fox and other predators such as badgers and dogs are a constant problem. Making sure the coop/run is predator-proof is my biggest job. Wiring the bottom of the run can be done very easily but affects the chickens' ability to scratch, so it is not something I favour. Instead, if you have a movable run, put skirts of wire around the edges and secure them with tent pegs. This works very well if you have a fixed permanent run. Concrete slabs (see left) are brilliant as they stop any predators digging underneath and also give you a path to walk around, which makes checking the perimeter easy and mud free!

Human hair is said to be a deterrent – hairdressers are usually happy to supply you with this. You can stuff the hair into old stockings and hang them from the run. Men's urine is also said to be hated by foxes, so you might want to encourage any males to pee around the outside of the coop/run. However, a gamekeeper friend informed me that although this might have been true years ago, foxes now associate it with food, particularly in urban areas. If you live near a zoo and can get hold of lion dung that is also supposed to work! There are also products on the market for scaring foxes, which use lights and/or sonic sounds that are not heard by the human ear.

Even if you think you have a fox-proofed home, change your routine occasionally. Let the chickens out later or put them to bed earlier. Sometimes leave a radio on, or place a solar garden light nearby, rattle cans hung to fences – something new will keep Mr Fox on his toes.

Electric fencing (see page 27) is used by many chicken keepers, including myself. This works brilliantly – but there is always a but! Maintenance is very important. In order for an electric fence to work correctly it needs to be set up properly and the manufacturer's instructions followed. The area directly under the electric fence needs to be clear of anything that could 'ground' the electric current out. Long grass should be regularly strimmed/cut and cleared away, and you need to check for anything that might have fallen on the wire every day. I check my fences daily with a current tester to make sure sufficient current is passing around the fence – when there is it is a very effective deterrent.

Most faults are simple to fix but when there is a problem your chickens are vulnerable. I have kept chickens for many years and have had my fair share of disasters, mainly due to Mr Fox – sometimes because of my own stupidity and other times due to the fox's cunning persistence. On one occasion I had been having my morning cup of tea watching my girls (45 stunning Buff Orpington pullets) that were coming on tremendously well. I was so proud having hatched and reared them, and because I was with them I had switched off the electric fence. I was sitting there happily watching them frolic and play, when the phone rang. I left them for all of 20 minutes while I chatted to a customer, and was in the office (barn), not more than 40 metres away, where they were just out of sight. The customer was enquiring about my birds, and when I finished speaking and started to walk back to the girls, I realised something was wrong.

A fox had jumped the fence and killed all but two of them, which I had to despatch because of their injuries. I strongly believe he was watching me, saw his opportunity and took it. So never let your guard down because this is when they strike.

On another occasion I had seen some signs of digging around my Brahmas' cage. I had meant to inspect it more closely but my children distracted me and I forgot – unfortunately the chickens paid the price that night with their lives. It would have been only a quick job as the fence just needed pegging down. If only I had done it!

Whatever your view is of foxes, it is your responsibility to protect your girls. It is not a good idea to feed foxes in your garden if you have chickens, which some of my customers do, as it really is asking for trouble. It is a fox's natural instinct to hunt and kill, even on a full stomach.

Remember that you can fox-proof your coop and run, and put up lots of deterrents, but if you don't close the door you've lost the battle. It might sound silly but, believe me, it happens regularly. I recall a morning when I noticed the three men who work on the farm walking briskly in my direction. This was very unusual, they don't usually move that quickly! 'This one needs you,' they said, and as I walked down the drive to investigate I found a customer coming up the drive crying. She told me that she had not closed her coop properly and had lost all her chickens to the fox.

Badgers will kill chickens. They are much more powerful than foxes and can cause terrible damage.

Seasonal care

Spring

In the spring, the weather is cheering up, the garden is coming to life and, for me, this is a magical time of year. The hard winter is behind us and at last, hopefully, a few months of good weather lay ahead. It is a busy time of year on the farm and early spring is spent mending fencing runs and generally tidying up. Now is a good time to do any maintenance on the chickens' run as the ground has thawed, and to do any re-seeding, as long as the frost appears to have finished.

The girls will be eager to get out and about. You should check the coop for any mites or damage and treat appropriately. This is when pests begin to breed, so clear away any debris that might become nesting sites. Be very aware that fox cubs are being born and soon they will be out hunting. I find foxes become really desperate towards the end of spring, when the cubs are weaned from their mother, but are not quite old enough to hunt for themselves.

Summer

In summer, with lovely long, hot days, life can be wonderful and it is very easy to think we can just sit back and enjoy it. Unfortunately these days are perfect for mites to breed and they really do breed rapidly if you don't keep a watchful eye out for them. Summer showers can cause warm, damp, humid conditions that in turn can be the perfect breeding ground for disease. There is little we can do about the weather, but routine jobs and tricks can make life a lot easier.

Check for mites, and if found treat them immediately and thoroughly (see page 124). Clean drinkers regularly as algae can form very quickly. If you wouldn't drink the water, replace it! Make sure the chickens have adequate shade – heat stroke can occur and they can dehydrate very quickly (see page 130). Small water bottles frozen overnight and placed in the run can be used by the chickens to lay on and cool down! Some love it, others don't. Some love to be sprayed with a fine mist of water.

Citronella is very good if sprayed around the coop to deter flies. Fly traps hung close by (but out of the reach of the chickens) work very well. Make sure you worm your girls regularly. Towards the end of summer and into the early autumn is when most chickens go through the annual moult, when they naturally lose their feathers and replace them with new ones. A chicken going through its first moult may only lose a few feathers but the second and subsequent moults can be quiet dramatic and alarming to a new chicken keeper. With pure breeds egg laying slows or stops when the chicken is

Seasonal tasks

SPRING
- Do any necessary maintenance of runs and coops
- Re-seed runs if needed
- Check for signs of rodents

SUMMER
- Check for mite
- Provide shade for chickens
- Hang citronella/fly traps
- Collect and freeze berries for use in the autumn and winter

moulting; with hybrids, depending on their age, laying may continue. The annual moult can take up to eight weeks to complete. If a chicken is still laying whilst going through the moult, it can put a huge drain on her. It takes a lot to stay healthy and lay lovely eggs as well as replace feathers. Giving her a tonic will help. Feathers are made of keratin, a fibrous protein, so adding this to the diet is very beneficial. Removing layer's pellets from her diet and replacing with corn may stop her laying but not always.

A daily fresh mix of oats, seaweed, cod liver oil, garlic and poultry tonic, with a little water, is a great help when fed in the early afternoon towards the end of summer and autumn. If you collect berries from the hedgerows the girls will love a few of these. For a cooling summer treat, freeze them in individual ice-cube trays and pop one or two into the run on hot, sunny days. If you collect lots, freeze some and bring them out in the dull winter months. The chickens will love you for it – just make sure they are defrosted.

Autumn

Some chickens may just be starting their annual moult, while some will be coming to the end (see page 114). It is a good idea to check all fencing and the coops for signs of wear and mend where appropriate. You don't want to be doing these jobs in the winter, when the days are shorter.

Your girls may have stopped laying, or laying sporadically. This is perfectly normal for older girls and pure breeds, but unless there is a very dramatic change in the weather spring-hatched hybrids should continue to lay. Quite a few poultry keepers will restock with a few young chickens at this time of year to see them through the winter months.

Commercial producers routinely use artificial lights in the coops, set to give a few extra hours in the morning and evening, but I prefer a more natural approach and believe my girls deserve a break, as nature intended. As the leaves start to fall you can collect them and place them in piles in the outside runs. This gives the girls a fresh place to scratch, and you can replenish them whenever you need to as they are free. If you have a particularly sunny day, take this opportunity to give the coop a thorough wash with disinfectant, but make sure it is completely dry by the evening.

If you can find an alternative coop, such as a Wendy house, garage or even a greenhouse, and put in suitable bedding you could also treat the coop with wood preserver, which is very beneficial. Also check that there are no leaks in the roof. When the house is completely dry and well aired, pop the chickens back in ready for the winter.

ABOVE: Hedgerow blackberries can be harvested and frozen to provide treats throughout the year.

Winter

During the winter you need to continue making checks on perimeter fences because, as the days become colder, predators will become more determined and will not miss an opportunity to get in. It is useful to coat hinges and locks with grease – there is nothing worse than trying to mend frozen hinges in the snow. You need to be particularly vigilant about rodents who are looking for spilt food and a place to live (see page 87).

A lot of people leave their chickens in one place over the winter and add shelter and windbreakers to make them happier. If you decide to do this, treat the rest of your ground with a covering of lime which will help reduce acidity in the soil. This is said to also help with internal parasites such as worms.

If you put up shelter over the run, try to make it so that you can roll it back or remove it on nice days. If there is any snow it is good to put tarpaulins over the runs. This enables you to get the snow off in the morning, and leave you with a snow-free run and roof; it has been known for roofs on runs to collapse when snow is not removed. In extreme weather conditions when the temperature plummets, coat the chickens' wattles and combs with petroleum jelly to prevent frost bite (see page 130). Water will need to be checked regularly when the weather is freezing. I have found it useful to replace my hanging drinkers with shallow dishes which are easier to empty of ice and to fill. If you make a warm meal for the girls, it will be greeted with great excitement and it makes you feel better about them being outside. Each day make a freshly mixed meal of boiled potato skins, bran, oats and cod liver oil.

If the weather is freezing for prolonged periods, it is beneficial to deep litter the chickens. Instead of cleaning out the old bedding, put a fresh layer on top and continue to do this until the weather breaks. Leaving the dirty bedding under the fresh will produce heat, which will rise and keep the girls warm. When the weather breaks a thorough clean will be necessary. This is a method used by a lot of farmers and small holders to keep their stock warmer.

AUTUMN
- Check for mites
- Give tonic to moulting girls
- Set up artificial lights to lengthen the day (this increases the egg yield)
- Give a maintenance check to runs and coops
- Collect fallen leaves for the runs
- Give a last big clean of the house and treat with wood preserver

WINTER
- Provide protection from the elements
- If weather is extreme, cover runs with tarpaulin and remove snow daily
- In freezing conditions replenish water frequently
- Make constant checks for predators as they will be desperate for food
- Administer poultry tonics
- Give Brewer's yeast for a few days to combat lack of sunlight

RIGHT: Hanging greens provides both entertainment and a workout for the girls.

Chicken behaviour

Once you have shared your life with chickens it is very difficult to see them as anything other than members of your family. Their behaviour is not dissimilar to our own – except for scratching for bugs!

Every flock of hens will have a top girl who calls all the shots. She will eat first, assume the right to have any bugs found and have the favoured perch position at night. She will also look out for her peers. As a flock, when needed, they will work together. It can be upsetting to watch the bottom girl do everything last and gaze wantonly at the worm that was not to be, but once the girls have established their place within the flock they will settle well and get on with it. A flock with all top girls would be a stressful, unhappy environment to endure. A balance will be struck, as in everyday life.

Scratching, dustbathing, preening and sunbathing are all normal undertakings for chickens and this is how they like to fill their days. They also argue and boss each other around and they like to explore, given the chance. Some people say chickens are stupid. They are not. They will remember where they can escape, for example, and the location of the vegetable patch, and when to cluck and look cute for treats. Chickens are also very astute at spotting signs of danger and will make alarm calls. Aeroplanes, however, freak them out and send them running for cover.

Laying their eggs is always an exciting part of the girls' day and evokes gregarious clucks of achievement, which always manages to make me smile. Once you start watching your hens the time flies by, their antics are hilarious.

Chickens use their eyes independently, and in daylight they have very good eye sight and can see different colours very well. In stark contrast they are almost blind at night, so that is the time to catch and handle your birds. It is much easier for you and the bird. They have a special ability to feel vibrations from the ground, which helps them when out foraging to sense predators or danger. If you walk up to your poultry house quietly at night you will immediately hear the warning calls of your girls as they feel you coming.

CHICKEN SAYINGS

Several common sayings we use daily come from our observations of chickens, including:

- To clip someone's wings
- Pecking order
- Cockiness
- To ruffle someone's feathers
- Nest egg
- Beady eyes
- To chicken out
- To take someone under your wing and
- To get in a flap

RIGHT: Hens enjoying a dust bath, which helps to keep their feathers in tip-top condition.

Studying cows, pigs and chickens can help an actor develop his character. James Dean

LEFT: A Poland cock leaping onto a stable door.

Chickens have a very good sense of hearing. You will notice the different calls they make from alarm calls to clucks of pleasure. Mother hen and her chicks are a joy to watch and she's very vocal as she teaches her youngsters what to eat and where to scratch.

If you have a cockerel in your flock you will notice his different calls, from cock-a-doodle-doo (i.e., look at me) to a low mumble when he finds a tasty morsel to eat.

Chickens develop their own routines. When you let them out first thing in the morning they are full of life, ready for a drink and something to eat. They then settle in for a day of meandering around looking for a tasty morsel to eat, dustbathing and preening their feathers. The occasional argument may develop if one of them finds a bug and doesn't eat it quickly enough. Napping is common if the girls are comfortable in their group. When you give them their treats in the afternoon, activity starts again. As they fill their crops for the night ahead, you might notice the lower-ranked girls going to bed first in a bid to get the best perch, only to lose their place when the higher-ranked girls go inside and make them move. They will only go to sleep when it is absolutely dark and silent and will wake at the slightest noise or vibration.

A chicken can peck another for no apparent reason other than to show that it is stronger. Problems can occur when insufficient numbers of chickens are kept together. A lady rings me constantly about her two girls as only one of them gets pecked. I have explained about the pecking order and how difficult it is to establish this with only two chickens. It really is best to keep a minimum of three hens together. One more doesn't make much difference to your chores or feed bill, but it does make a difference to their behaviour.

Some breeds of chickens can fly much better than others, but when we say fly it is more of a gliding jump. The longest recorded flight of a chicken lasted for 13 seconds. They are not capable of sustained flight.

BROODY HENS

A broody is a chicken that wants to be a mum. Certain breeds are more likely to go broody than others, for example Pekins and Silkies. Hybrids are less likely to become broody as their main aim is to lay lots of eggs, but it is not unheard of.

I receive lovely phone calls about broody hens that always make me smile. Although tyrannosaurus rex is a very distant relative of the chicken, a chicken has no teeth and when handled correctly can do you no harm. However, when I hear some people's descriptions of their broody hens' behaviour it resembles that of a rabid dog more than a protective mother. Yes, they are moody and determined, but this is behaviour to be admired not feared.

A broody hen will sit in her nest box and not move. Her breast will be hot and she may have plucked out some feathers to make her nest. If you approach she will fluff up all her feathers in a bid to make herself look bigger, she will almost growl at you as a warning and will peck her hardest. Just remember a peck is normally nothing more than a pinch.

If the eggs are fertile let her remain on the nest and hatch her eggs. Eggs take 21 days to hatch (see page 70). But if you haven't got a cockerel with your girls, any eggs will not be fertile and won't hatch.

You can often buy fertile eggs from breeders or on the internet. If using the internet remember an egg that goes through the post might not hatch. Also when hatching yourself you will get cockerels and, in my experience, lots of them.

If your broody hen is sitting on infertile eggs you need to remove them and try to break the habit. Broody hens will need removing from the nest twice daily. Just pick her up and pop her by the food and water. She will usually defecate and trot back into her nest box. Do this in the morning and again in the afternoon when you put the corn down.

If you want to break the hen's broodiness you need to cool her down. This can be done by placing her for a while in a broody box – this normally has a wire floor so the air can circulate and make her cool. Another method on a dry, warm day is to place her bottom in a bowl of cold water and leave her in the run, blocking off the nest box. Placing small, frozen water bottles in the nest box sometimes works. If a chicken is very determined it can take quite a while for her to stop, so persevere. A broody hen that is left to hatch her eggs will be a sitting target for mites and will need de-miting with powder regularly.

RIGHT: A broody frizzle Peking hen looking rather unhappy after having been removed from her nest for lunch.

Eggs

ABOVE: The best job of the day –
collecting the fresh eggs.

THE ANATOMY OF AN EGG

An egg is made up of three parts – the yolk, the albumen (the white of the egg), and the shell. A female chick starts life with two ovaries, but only the left one matures, while the other one withers away or atrophies. The ovary has a clump of undeveloped ova or 'yolks', which are the yellow part of the egg. When one is released the process of making an egg begins.

The ovum travels from the ovary down the tube (oviduct) to the magnum, where the albumen (white of the egg) is added. The partially formed egg is then passed further down by squeezing movements to the isthmus, where the membrane (which you notice when peeling a boiled egg) is added.

It then passes to the uterus, where the shell is formed; this can take up to 20 hours. When the egg is laid, the 'bloom' (the moist coating which can just be seen as the egg is laid, and which helps to keep out bacteria) dries out. It takes around 24 hours for an egg to be made. In point-of-lay chickens, which are just reaching maturity, it is not unusual for two ova to travel down the magnum at the same time, and this will give you a double yolker. It is not unheard of to have triple yolkers, which causes great excitement with my children and the dilemma of who's going to eat it! The record of nine yolks in one egg has been recorded. These incidents decline and settle as the chickens mature. At point of lay they can also pass soft eggs (eggs without a shell). If this happens it is again usually because their bodies are young and the egg passes through too quickly. Only if this happens frequently is there a problem, in which case consult a vet.

So not only are your girls busy all day scratching and running around, inside their bodies they are always on the go as well. If you are lucky enough to watch one of them lay an egg, it is truly fascinating. I have a wonderful memory of my daughter's face when she had watched her chicken lay its egg. She came to me with such a sparkle in her eyes, clutching the warm egg. Chicken keeping is such a simple hobby, but it has truly enriched my family's lives.

Point-of-lay (P.O.L) pullets usually start laying at about 19–21 weeks, depending on the time of year. In the warmer months they can start sooner; towards the end of the year it will usually be later. Indications that your chicken is close to lay are a lovely red comb in some breeds, such as the Leghorns, and in Rhode Island Reds their legs will be yellow. Another indicator is the gap between its pelvic bones. A hen that is laying will have a gap the width of about three fingers.

Before a female chicken lays her first egg she is called a pullet, after that she is a hen.

Egg colour is often related to the colour of chicken's ears - red ears generally produce brown eggs and white ears white eggs.

It's a very exciting time when you get your first chickens and are waiting for them to lay. I can still remember when I was a small child waking up first to make sure I could collect the eggs. The disappointment of discovering there were no eggs was immense! But then there was the reignited excitement the next day when I was going to check again. It is a magical time not only for the chickens, but for both the adults and children sharing the experience. When your chickens start laying, their eggs can be quite small. They can also be empty, with no contents in the shell. Don't worry as this will settle down and the eggs will get bigger as they mature.

I had a lovely email from a family who had purchased their chickens from me. One of them had laid an egg the next day and the mother said that it was like Christmas morning in the kitchen – everyone was so excited as they boiled the precious egg and cut it into four pieces so they could all have a taste. 'It was such a wonderful experience that has brought us so much joy; I wish I had done it years ago,' she wrote.

You need to remember that even if you buy a chicken who is already laying, sometimes the journey home and stress of living in a new coop may put her off laying for a short while. Just be patient and it will happen again.

WHAT'S IN AN EGG

Eggs are mostly water – around 75 per cent – and are composed of three main parts:

- Shell
- Egg white
- Egg yolk

The shell accounts for around 11 per cent of the total weight and has tiny pores that not only allow the developing chick to obtain oxygen but also allow water and carbon dioxide to escape. The shell is generally strong, but older hens tend to produce weaker shells. The colour varies according to the breed of the bird.

The colour of the egg yolk is determined by the diet of the hen and is due to the presence of carotenes and any colourings added to its feed. The nutritional value of the egg is not affected by the colour of the yolk.

shell membranes

chalazae

shell

thin albumen

yolk

thick albumen

air cell

How long do chickens lay for?

How long a chicken will lay for depends on the breed. Pure breeds tend to lay seasonally, meaning that most hens come into lay in the spring, when the weather begins to get better and the days slightly longer. They continue to lay until the end of autumn, when they moult, and usually have a rest throughout the winter months. As pure breeds have natural breaks in laying, they can lay for years quite happily. My chicken Henrietta laid until she was eight and a half years old, although admittedly in later years not as well as she did in her youth.

Hybrids will lay consistently for the first 12 months. Commercially they are called 'spent' hens by the age of 18 months when they are usually despatched or re-homed, and then a new group of hens is brought in. In a domestic flock they will begin to lay again after a break and will continue to do so for quite a few years. Most people find that killing their hens when the eggs drop off is acceptable, while others choose to keep them even though their laying is very sporadic. The choice is yours. You can always introduce new hens at a later date to give you more eggs (see page 66).

How fresh are your eggs?

Most 'fresh' eggs bought in a supermarket are three weeks old at the very least. To check how fresh an egg is, place it in a bowl of cold water. If it lays almost flat on the bottom, it is very fresh. If it tilts slightly fat-end up, it is older. The higher it tilts, the older the egg. If it floats on top of the water it is very old and shouldn't be used. Be careful not to break it, if you do it is guaranteed to smell horrible.

Storing your eggs

Eggs in my house are never around for long and are kept in a ceramic egg tray on the work surface in the kitchen. I am told that ideally they should be stored in the fridge in their carton. I think fresh eggs are fine kept in the kitchen, where I like to see them. Several friends put their eggs on view in baskets – with colours that range from white to light brown, chocolate brown and blue – they are too beautiful to hide in the fridge.

> ### THE NUTRITIONAL CONTENT OF EGG YOLK
>
> - 16.5 per cent protein
> - 33 per cent fat
> - 50 per cent water
> - Vitamins A, D, E and K
> - Mineral elements, including iron
> - Lecithin (an emulsifier)

LEFT: Examples of the different size eggs a chicken can lay. The bottom right egg is a pullet egg. (Do not despair they do get bigger!)

It takes about 24 to 26 hours for a chicken to lay an egg, so expect one egg a day, although apparently the record is for a chicken who laid seven eggs in one day.

Do you need a cockerel?

One of the questions I'm most frequently asked is, 'Do I need a cockerel to make my girls lay eggs?' The answer is no, your chickens will lay eggs whether you have a cockerel or not. You only need a cockerel if you wish to breed from your hens.

However, if you can have a cockerel it makes a lovely addition to your flock. Males tend to have a calming influence on the girls, and quickly break up most squabbles. They are fascinating to watch. I have had some lovely cockerels. Brigadier, a huge, gold Brahma, used to look after his hens like an English gent, calling them when he found a worm. His presence within the flock had a calming effect and arguments were nonexistent. He was always last to bed, almost counting his girls in. On the other side of the coin there was Rocky, a Blue Cochin, who went from a lovely boy to a cockerel from hell overnight, attacking anyone who went near him and inflicting some pretty awful injuries on me. Sadly, he had to go, as no one wanted to go near him. Most cockerels are friendly and become protective of their girls during the breeding season from early spring to late summer, and cause no problems.

A cockerel's spurs can become very big, so it is a good idea to keep them trimmed. But be careful when you do this as it is very easy to make them bleed.

Introducing a cockerel to your flock is usually straightforward as long as it is mature – aged about one year old is best. If it is too young it will be bullied by the hens. The only problem you might encounter towards the end of the breeding season is bare patches on the backs of the girls. This is caused by the male treading (mating) his girls. It usually only occurs when a cockerel has two or three girls. You can spot his favourite girl as she will be the baldest. Her feathers will re-grow in three to five weeks. Saddles can be purchased via the internet to protect their skin if it gets too bad. These come in different sizes to accommodate most breeds. If you use them you will need to check regularly for mites and infections.

Storing eggs

It is often recommended that eggs are stored in a refrigerator and, because they are porous, away from strong smelling foods. If you do keep them in a fridge, they will remain fresh for up to 28 days. If you prefer not to do so, they will last for 21 days.

As eggs age, bacteria enters through the shell. Water moves from the white to the yolk, the yolk linings weaken, the white becomes thinner and the egg decomposes as the bacteria contaminates the contents. Outwardly the shell begins to look dry as it loses moisture.

RIGHT: A stunning Brahma cock overlooking his land.

A young male chicken is called a cockerel; when he is fully grown (about a year) he is known as a cock.

Health and wellbeing

Anatomy

It is not necessary to know everything about the anatomy of chickens in order to keep them, but it can be useful to have a small understanding of how things work so you can recognise early on if there is a problem. For instance, when an egg is laid without a shell. It's also useful to know where their ears are – most people do not realise they have them. Knowing how chickens eat will enable you to keep an eye on their diet, and realising what a cecal dropping is can stop you panicking for no reason. All of which will hopefully help you relax and enjoy your girls without worrying.

RIGHT: A Rhode Island Red Blacktail hybrid free ranging, looking very happy and healthy.

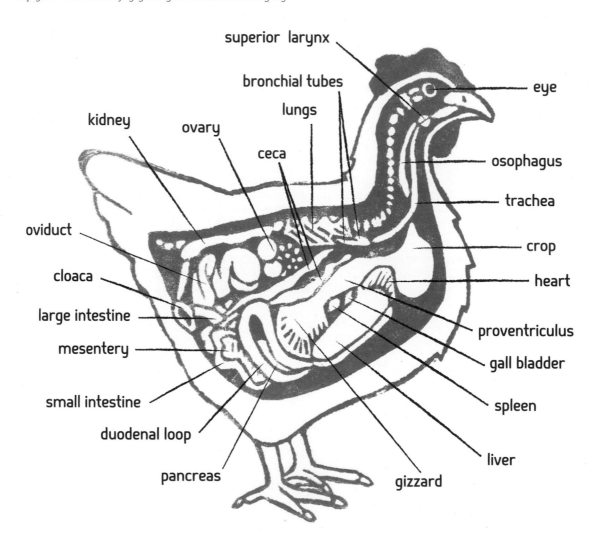

superior larynx

bronchial tubes

lungs

kidney

ovary

ceca

eye

osophagus

trachea

crop

heart

oviduct

cloaca

large intestine

mesentery

proventriculus

gall bladder

small intestine

duodenal loop

spleen

pancreas

liver

gizzard

COMBS, BEAKS, EYES, WATTLES AND EARS

• **Combs:** There are many types of combs – they come in all shapes and sizes. They are not all red, some are purple. What shape and size they are depends on the breed and sex of the chicken. Cockerels tend to have larger combs which are used to attract females. Their main purpose is to help cool the birds in hot weather.

Chickens do not sweat; instead they circulate blood to their combs and wattles, where the body heat radiates and cools the chicken. The seven common comb types are: butter cup, cushion, peg, rose, single, strawberry, and V-shaped.

In very cold weather the chickens can suffer from frost bite. This can be avoided by applying petroleum jelly (see page 130).

• **Beaks** are used for picking up food, and should be a natural shape. De-beaking a bird is a common practice in commercially kept birds, as it stops them being able to feather peck each other. It also stops them from foraging naturally as they are unable to peck the tips of grass. It must be very difficult for them to pick up any small object – rather like trying to eat grapes with chop sticks, and is not a practice I agree with.

• **Eyes:** Chickens have very good eyesight and see in colour. Their eyes work independently.

• **Wattles** hang below the beak and are usually bigger on the male. They are used to help cool the chickens in hot weather.

• **Ears** are small holes covered by feathers. There is a link between the colour of the earlobe and the colour of the egg laid. Chickens with white earlobes always lay white eggs. Those with red ones usually lay brown eggs.

ABOVE: The bright eye of a healthy chicken. OPPOSITE TOP LEFT: The comb of a white Silkie cockerel. TOP RIGHT: A normal chicken beak. BOTTOM LEFT: A chicken's ear. BOTTOM RIGHT: The blue skin of a Silkie.

Feathers

OPPOSITE: examples of different feathers. TOP LEFT: Speckled. TOP RIGHT: Single lacing. BOTTOM LEFT: Double lacing. BOTTOM RIGHT: Millefleur.

Feathers repel rain a little, keep chickens warm and protect their skin from the sun. They also act as a cushion to help protect the body from injuries. The different patterns include: barring, a striped pattern of light and dark alternating horizontally; lacing, where the edge of the feather is a different colour to the middle; spangling, where there is a splodge of colour on the tip of each feather; double lacing, which is very pretty and has two areas of dark colour; cuckoo, where stripes are not as regular and may blend into each other and pencilling, very fine stripes which follow the shape of the feather.

Chickens either have hard or soft feathers, depending on the breed. Hard feathers grow very close together and cover the body tightly. More fluffy, soft feathers attract many unwanted visitors (see pages 122–5). Cockerels have different feathers to the hens – they are longer, more pointed and shinier. Some breeds of chickens even have feathered legs. Birds naturally maintain their own feathers very well and spend a great deal of time preening. They have a gland at the base of their tail that produces oil, which they transfer to the feathers by their beak.

Feathers are moulted and replaced annually, usually after the breeding season. It may happen over the entire body all at once, or in a specific pattern, starting at the head and neck, followed by the breast and body, and then the wing and tail. It's a stressful time and your girls may become lethargic. The process takes between three and four weeks, but can last up to two months. During this time egg laying can cease, though this occurs less frequently with hybrids. All chickens benefit from poultry tonic to help them re-grow their feathers and lay eggs.

WING CLIPPING

Some breeds of chicken are rather good flyers, and are capable of flying onto a shed roof. Others are great escape artists. I advise clipping your chickens' wings when you first get them, as they can be stressed and nervous when introduced to their new home. You only clip the left wing, as the active ovary makes the bird slightly heavier on this side. This unbalances them, making them less agile and able to fly. If done correctly it is neither painful nor cruel, but is rather like cutting your nails.

You need to cut the ten flight (primary, quill) feathers (see left). Make sure you don't cut any lower than the tip of the other feathers, and never cut the secondary feathers, as the girls use these to keep themselves warm. Wing clipping is a temporary measure as the feathers will re-grow. A friend of mine bought some chickens from me and I clipped her girls' wings for her. She then went home and cut the other wings on all of them because she thought I had forgotten. A week later I went to visit her and was greeted by her girls getting over her fence and running up to us for biscuits. I couldn't believe it until she sheepishly owned up to what she'd done. We laughed about it as there was nothing we could do, other than raise the fence. The downside of wing clipping is that it does affect the chickens' ability to get away from predators, so only do it if you really have to.

DO ALL CHICKENS NEED WING CLIPPING?

I would recommend that all point-of-lay (P.O.L) hybrids are clipped initially, unless kept in a fully enclosed run, and certainly all White Leghorns and Skylines.

LEFT: The flight feathers of a chicken's wing before clipping.
RIGHT: Clipping the flight feathers.

Legs, spurs and toes

Chickens' legs are covered in scales, which can be troubled by mites (see pages 124–126). Their thighs are covered in feathers. Spurs grow only on cockerels and they use them to protect themselves and their girls. A cockerel's spurs can become very big. Even if he is friendly towards you, it is best to keep them trimmed as they can inflict some awful injuries on the hens when he is treading (mating) them. This needs to be done with a good pair of clippers, as the spurs are extremely strong. You just need to take off the sharp tip, not the whole thing as it will bleed profusely.

Chickens normally have four toes, but some breeds, such as the Silkies, have five. They can injure their toes but normally they heal quickly with the right treatment. On the end of each toe are nails which they can lose or break (see page 130). Most chickens don't need to have their nails clipped because they are constantly scratching, which keeps them neat. But there are exceptions, light, small breeds sometimes need attention. It is very easy to do but you need to be careful not to clip too much. You can normally see the little blood vessels to avoid. Just take off the tip. If you take off too much they will bleed a lot, so spray the area with disinfectant and put on a little smear of petroleum jelly. Then keep the chicken separate from the flock so you can keep an eye on her. The bleeding should stop quite quickly.

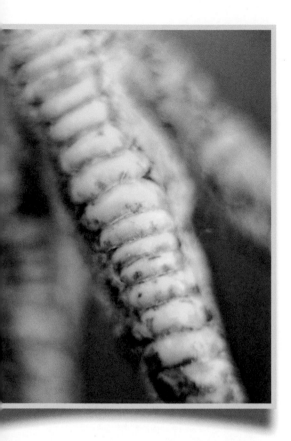

ABOVE: Close-up of a healthy chicken toe.

How chickens breathe

Chickens have two lungs that are connected to the trachea (windpipe) like us but that is where the similarities end. Their lungs are fixed to their ribs, so they do not expand. Birds pass air through their lungs using the air sacs that only birds have. Air sacs are small, balloon-like structures that store air inside and act like bellows; this means that the lungs can constantly have air flowing through them. Since chickens depend on the movement of the chest and rib cage to breathe, holding a chicken too tightly can suffocate it.

Chickens have nine air sacs altogether: two surrounding the intestines (abdominal), two at the neck (cervical), one at the shoulder (interclavicular), two beneath the lungs (thoracic anterior) and two behind the lungs (thoracic posterior). Chickens' lungs get fresh air during exhalation (breathing out) and inhalation (breathing in) because it is the air sacs that swell and shrink, and not the lungs. Air flows in through the beak, and half of it goes to the lungs and anterior air sacs. The air from the lungs and anterior sacs then goes out of the mouth, but the air from the posterior air sacs goes back through the lungs. At any time, air can be flowing in and out of the lungs and stored in the air sacs. Chickens also have some hollow bones called pneumatic bones which aid the respiratory system.

Digestive system

When a chicken eats its food goes down the oesophagus. It stops at the crop, where it gets softened. The food then goes through the proventriculus, where the digestive acids are added, and passes down into the gizzard, which has two sets of strong muscles to grind and mash – this is where grit helps. (see page 36). Then it goes through the intestines, where the nutrients and water are absorbed. Food waste exits through the vent. When a chicken lays an egg, the vagina folds over to let the egg out through the vent, so that it doesn't come into contact with any faeces.

The crop is an internal pouch at the base of the chicken's neck by the breast and is used to store food and water. You will notice it towards the end of the day as it fills with food. A chicken uses its tongue to push food to the back of its throat, but this doesn't work very well with water, so when it drinks, it tilts its head upwards and opens and closes its mouth quickly, which allows the liquid to run into its crop.

Chickens have two blind pouches where the small and large intestines join. These empty a few times a day and produce horrid-smelling faeces called cecal droppings. They are usually a mustard colour. These let you know that your chicken's digestive system is working correctly. Normal chicken droppings should be firm and brown with a white cap, which is the urine. Yellow and foamy droppings, or those with blood in, can be a sign of coccidiosis (see page 128). White runny droppings can be an indication of worms and green of a severe infestation (or that the bird has eaten too many greens).

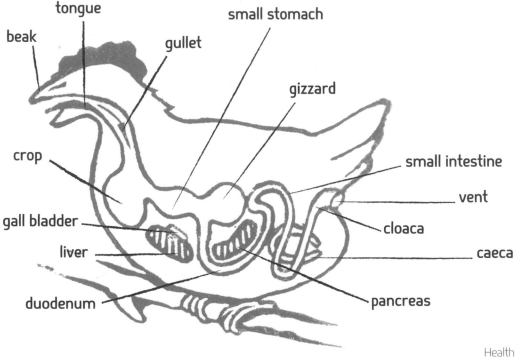

General health

Knowing what is normal for your birds is an important part of chicken keeping, as spotting signs of illness early can be crucial. In the wild they are prey, and being healthy is necessary for their survival. Chickens are very good at appearing well when you're around; if your girl is exhibiting signs of illness you can be sure she's been ill for a while. A typical stance of a poorly chicken is with its head down, hunched and looking depressed. If you spot these signs, separating your chicken is imperative otherwise she will be pecked and bullied by the others. This will also protect the others from whatever is wrong. If another house is not available, a temporary solution might be to put her in a box indoors with food and water.

I am not a vet and the remedies I suggest are from my own experience. If you are unsure, ring the person you got your birds from as they will usually be happy to help and advise. If symptoms persist and do not respond to treatment, I suggest you contact a vet – but bear in mind that not all vets deal with chickens regularly enough to warrant a visit. If you obtained your chicken from a good supplier they will have their own poultry vet and will be happy to give you the phone number. This is another reason why buying from a good supplier is advantageous.

In my experience, natural treatments really do help if administered on a regular basis in the correct amounts. It must be noted that these do not replace wormers, coccidiostats and veterinary help, and will not make up for inadequate food or water you wouldn't drink yourself, insufficient space or poor living conditions.

The mention of bird flu makes every chicken keeper freeze with fear. Remember, we hear a lot about it, but it is really confined to poorer parts of Asia where people live in the same room as their chickens, and you are very unlikely to encounter it. However, the signs to look for are: their combs and wattles may become blue; the hens will look ill and stand around hunched – and death comes very quickly. It is a notifiable disease. If you have large numbers of birds dying, contact a vet immediately. Some medication prescribed by vets can affect the eggs, so during treatment they should not be eaten. A vet will be able to advise you on this.

Despite its name, chickens aren't affected by chickenpox – it is solely a human disease.

RIGHT: A group of hybrid hens taking refuge from the rain.

ABOVE: Lice eggs are normally found at the base of feathers, particularly around the vent.

A-Z of parasites

Flukes

These parasitic worms are leaf-shaped and flat. Wild birds can introduce them to your chickens but they rarely cause problems if their living conditions are good. They can affect the eyes, skin and the oviduct, causing it to swell and sometimes burst, which will kill the chicken. Symptoms include droopiness, chalky white droppings and weight loss.

Treatment involves keeping your chickens away from ponds, lakes or boggy land because the fluke needs two hosts. By controlling where the chickens graze you can control the host and eliminate the problem.

Gapeworm

This is a respiratory problem as the adult worm lives in the trachea (windpipe). Symptoms are coughing, gasping, yawning regularly, throwing the head back with the mouth open, shaking of the head and loss of energy. When you hold your chickens you'll hear a gurgling sound. If the infestation is severe and left untreated the birds will suffocate and die. As with all worms (see page 127), land rotation is extremely important. The worms' eggs take up to two weeks to hatch. Treatment involves administering an appropriate wormer.

Infectious bronchitis

Otherwise known as I.B., this has very similar symptoms to mycoplasma, but, unlike mycoplasma, will infect the whole flock within a few days. Symptoms are coughing and wheezing. If left untreated it can also affect the nervous system and the chickens will have droopy wings and a twisted neck. In mature hens droppings may become green and they may start to lay soft eggs or not lay at all. Most hybrid chickens are vaccinated for I.B. Birds that recover from I.B. are immune from getting it again but will always be carriers.

Treatment involves keeping the infected birds warm and looking for any secondary infection that may need antibiotic treatment. Vitamin supplements added to their water helps. There is no known risk to humans.

Lice

These are an external parasite that is species specific, meaning they only like birds. They have a life cycle (egg to mature louse) of approximately three weeks. They spend their entire life on the chickens and will cause great damage to their feathers and their wellbeing. Signs of a louse infection are: the chickens become restless, and scratch and preen continuously; bald patches may appear, particularly around the vent, where the birds will preen away their feathers. Egg production may be affected as the chickens become stressed. If you examine your birds you will see straw-coloured louse scurrying around. They can move very quickly. They lay their eggs, which are white in colour, in clusters at the base of the feathers (see picture on opposite page). The eggs will hatch after four to seven days, depending on weather conditions, so treatment should be repeated as necessary.

Treatment is normally done by powdering or spraying the chickens regularly to break the cycle. The lice products are designed just for the animals; there is no need to treat the house because the lice live only on the chicken, making this easier to deal with than mites.

Mycoplasma gallisepticum

Commonly known as 'myco' in the chicken world, this is a common problem that a lot of chicken keepers experience. Being an opportunist organism means that it strikes when your birds are under the weather or stressed. The stress trigger can be something as simple as moving them to a new house, introducing new girls to the flock or sudden changes in the weather. Symptoms include runny 'bubbly' eyes, nasal discharge, coughing, sneezing, and, in severe cases, swollen eyes.

Wild birds can carry the disease and can easily infect a free-range flock. It is important to make the water and feed as inaccessible to wild birds as possible.

Unfortunately some chickens do die from this, while some will recover but have similar bouts throughout their lives, and others won't become ill but will be carriers, which means that they show no signs of the illness but can infect new birds that you introduce to your flock.

Treatment is usually a broad-spectrum antibiotic prescribed only by your vet. I cannot emphasise enough the need to obtain treatment quickly. There are also natural remedies that you can try. Two to three garlic cloves peeled and crushed, and placed in the chickens' drinker regularly can help as a prevention. It helps to keep the birds' immune system strong, to fight diseases. Apple cider vinegar added to their water helps with respiratory problems, as it seems to break down the mucus which can then be expelled more easily by the bird. Buy it from a pet shop or health food store. The apple cider vinegar sold in the supermarket is processed, and not as efficient as the pure versions that are sold for animals.

If your flock is infected with myco and you choose to cull it and start again, you will need to thoroughly disinfect your house, feeders and drinkers and leave them for at least two weeks before introducing new chickens. If possible move the house on to fresh ground. Mycoplasma gallisepticum poses no risk to humans.

Northern fowl mite

This is probably the worst problem of all as the mite spends its entire life on the chicken, which can do a lot of damage in a short time. You may see the mites crawling both on the chickens' eggs and on the chickens themselves. In severe infestations this causes dark scabs, especially around the vent Treatment is as for red mite (see opposite), but with more attention given to the chicken. I have found that petroleum jelly rubbed over the scabs helps after treatment.

Ticks

If your birds are free range on long grass, ticks might attach themselves. However, this is extremely rare. Removal of the tick needs to be done carefully. If you try to pull them off the head is often left in the chicken, which will become infected.

Treatment: A quick spray with surgical spirit will normally make a tick drop off. You can also buy tick removers.

Red mite (Dermanyssus gallinae)

This is a common problem for nearly all poultry keepers, but it is classed as an inconvenience rather than a disaster as it can be dealt with. How we perceive problems can have an effect on how we deal with them. If you find the mites, don't panic– stay calm and understand what you need to do. Remember red mites do not live on humans, they are host specific, which means that they live and reproduce with the chickens. In very severe infestations they may crawl on you, but a shower and change of clothes will remove them. Don't scream, jump around and cry, as this will not get rid of the critters (I have tried!).

A hysterical lady was once in my yard – she had discovered red mites in her chicken coop. After a chat we soon got the problem into perspective and had a logical action plan drawn up for her to deal with them. She went away with a few products, ready to begin her battle. Two weeks later she had won and now has a very good knowledge of what to look for and what to do next time!

The first part of your plan to deal with them is to understand them and their life cycle. They are external, temporary parasites that live in dark crevices and cracks in the chicken coop, particularly around the ends of perches. They feed on the chickens for between one and two hours at night, although if they are brought in by wild birds some must live on them during the day. They can live for up to eight months with no host to feed on.

The eggs are very small, oval and pearly white. They are laid by the female adult mite in dark cracks and crevices inside the coop and nest box. They usually target the area around the perches first. The eggs will hatch within two to three days in warm weather. Once the eggs hatch they become larvae, which have six legs and are red in colour like an adult mite. At this stage they do not feed on the chickens. They take a day to develop into protonymphs, which have eight legs like an adult mite but are smaller in size. They do, however, start feeding on the chickens and after about two days they become deutonymphs. These are larger and after a meal of the chickens' blood they turn into adult mites. Adult red mites can reproduce and do so rapidly.

Signs to look for are ash-like dust, found in cracks; your chickens seeming generally unhappy with a loss of condition; the hens stopping laying; their reluctance to go to bed at night; and spots of blood on their eggs.

One tip is to hang a small, white cloth in the coop at night and check for mites in the morning. Another is to put double-sided tape on the bottom of the perch, which will making spotting them easier and will trap them.

If you catch red mite early, it is much easier and less daunting to deal with than a bad infestation. It doesn't take long to look for them every day and in the long run could save you a lot of work. Remember, red mites are silent killers, so do check regularly, especially in the warmer months.

Treating your flock and coops for mites

Natural treatments to deal with red mites include Diatom Powder (Diatomaceous Earth), which is made of micro skeletons of fossilised remains, which have sharp edges that cut the red mites' waxy coating, making them dry up and die. It is very effective and available over the internet in a puffer bottle that works very well. It is thought that red mites don't like the smell that comes off the chicken's skin, or the taste of their blood, when they've had garlic. I crush a clove of garlic and add it to their water regularly, as it is very beneficial to chickens too. You can also buy garlic granules from pet shops that can be added to the feed. It doesn't make the eggs taste of garlic. Tea tree oil and citronella, which are sold to prevent mosquitoes, and eucalyptus, cedar wood oils and lemongrass are all said to help repel red mites, although I have not tried these.

When using natural remedies, it is usual to use more than one product at a time, such as Aiatoms and garlic together. Remember you need to break the mites' life cycle, so regular treatments are needed.

Chemical treatments contain the insecticide permethrin, which attacks the mites' nervous system. There is a huge choice of chemical products on the market. Most come in powder form and remain active for some time, but they will need to be used regularly. Some sprays work as a repellent, others by washing the outer wax off the mites, which causes them to dehydrate and die. There are smoke bombs containing permethrin that work very well for treating the coop as the smoke is able to penetrate most crevices.

BELOW: A group of red mites (*Dermanyssus gallinae*) breeding well in a crevice.

Frontline and Ivermectin are anti-parasite medications that are not licensed for poultry, but vets will prescribe them under their clinical judgment. They will eradicate the mites that bite your chickens, but will protect them only for a short while and not get rid of the problem. So you need to use them in conjunction with other treatments.

Eliminating mites from your hen house

Other methods of eradication include painting the house with creosote – this has to be the old-fashioned type to deal with red mites, not the creosote substitutes sold in DIY stores. Paraffin painted into cracks is also said to work. You have to remember that both are very smelly and, once painted, houses should be aired for a couple of weeks before you introduce the chickens again.

Blow torches work well as they incinerate the mites. You do, however, need to be extremely careful not to set your coop on fire. Pressure washes also work but do not kill the mites, and only wash them away. And steamers used for removing wallpaper can also be used towards the end of the year to deep clean all houses. Smearing petroleum jelly mixed with paraffin into all the cracks will smother the mites. Double-sided sticky tape stuck to the underneath of the perches will trap them going for a feed. Cola poured into cracks is said to dissolve the waxy layer of the mite, making it dry out and die, a sticky solution I have not tried!

Breaking the cycle

Whatever method you decide to use, if you have an infestation my advice is to use a couple of products to break the cycle, do not stop treatment too early, and routinely dust your chickens and their houses, as prevention is better than cure. In order to get the best out of the products you are using, make sure you thoroughly clean the house first of all bedding. If possible, take out any removable parts such as perches, floors and nest boxes. If you have a felt roof it is advisable to remove this as I have not found any product that can penetrate it, which makes it the perfect place for red mites to breed and thrive. Replace the roof with a corrugated or treated wood roof. If using sprays, make sure the house is dry before letting the chickens back in.

Scaly leg mites

These live on the birds' legs. They are tiny, and burrow in under the skin – they spend their entire life on the chicken. Treatment needs to be methodical and done thoroughly. Moving the birds on to fresh ground helps. Crusty legs are a sign that your girls are infected.

Treatment involves washing the legs gently and treating them with the appropriate spray. There are many of these on the market. I favour the

If you have no knowledge of chickens and are unconfident it is always best to get advice from a poultry vet. Certainly if your chicken is:
- badly injured
- showing signs of distress
- hunched with no appetite
- unable to walk
- has a problem that is not getting better

old-fashioned method of using surgical spirit on the legs and regularly smothering them in petroleum jelly. Scaly leg mites are very contagious, so don't introduce any new birds into the flock until the problem is resolved.

Worms

Chickens can pick up worms from the land, so land management, or rotation (moving the position of your coop) is very important in order to break the life cycle. Some worms can complete their life cycle with a chicken – i.e. the worm lives in the chicken – which defecates its eggs, which then re-infest the chicken. Some worms need another host in turn – the worm lives in the chicken, which defecates its eggs, which are then eaten by snails/slugs/beetles and earthworms, etc. The chicken then eats the infected bugs and the cycle begins again. Wild birds can host worms. This can be how your chickens become infected in the first place.

If no chickens are on the land for a while, there is no host for the worms to live in and so the cycle is broken. You can apply lime to the soil, which helps to deter pests, but removing the birds is the best way to break the cycle. Treatment: Wormers are available to buy and easy to use; there are chemical ones and herbal wormers that work by making the inside of the bird uninhabitable. Signs that your birds are suffering from worms include: pale combs, gradual weight loss, generally dull appearance, lethargy, their laying may be affected, and diarrhoea, which leads to a dirty bottom.

Farming lore suggests the best time to treat worms is during a full moon, when they are supposed to be at their most active.

ABOVE: The toes of a chicken. Inspect these regularly for signs of damage.

A-Z of other health problems

Chickens can get up to mischief and the odd scratch is unavoidable. Any wounds need to be cleaned thoroughly with diluted iodine, then dried and covered with antiseptic spray. You'll need to keep the chicken separate while it heals, as the others will almost certainly peck at it. In the summer, watch out for flies, as you don't want them bothering the wound. If you notice pus as it heals, draw this out by applying petroleum jelly mixed with garlic. If you are in any doubt, seek help from a vet.

Bumblefoot

This is caused when a cut heals but pus builds up underneath, so check any bird that is limping. Causes can be unsuitable perches, as well as scratching or treading on something sharp. Treat by cleaning the foot with warm water and applying iodine, which is available from most pet shops and chemists. If the skin has healed and pus is still trapped apply magnesium sulphate paste (available from chemists). Sometimes it will need lancing by a vet.

Coccidiosis

This is one of the most common protozoal diseases found in poultry and can occur even in healthy flocks. Gradual exposure usually allows chickens to become resistant to it, so by maturity most are immune. Outbreaks are normally most severe in young birds aged 3–6 weeks and they will need immediate treatment. Poor sanitation, overcrowding or stress (caused by a change in food or housing being transported or a drastic change in weather conditions) can all lead to an outbreak.

Crop impaction

This happens if birds are fed cut grass, have access to hay or eat something that can cause an obstruction and results in a very swollen, hard crop. When birds eat grass or greens naturally they peck off small bits, but when you feed them it and do not hang the food up or anchor it the birds will eat the whole thing. This can then form a ball. Your chicken will look miserable and stop eating because there is no room in the crop, and, if left untreated, the bird will dehydrate and starve to death. To try to move the blockage, place some olive oil into the bird's mouth and gently massage the crop to loosen and move the impaction. Hang the girl upside down and, with gentle downward movements from the crop to the beak, help her regurgitate the blockage. Only do this for very short periods, as she will find it difficult to breathe. If this fails, add a little more olive oil to soften the crop and wait to see if the impaction moves the other way. If the bird is still unwell and not eating after 24 hours seek the help of a vet.

Dehydration

If a hen goes without water for a couple of days, she may go into a moult and then lay badly for the rest of her life. Signs to watch out for are combs and wattles that are shrivelled and bluish and legs that look bony. Rehydration salts can help if added to the water and are available from chemists.

Ear mites

Chickens' ears are commonly affected by mites that go into the ear cavities and start infections – the symptoms will be yellow pus and a horrid smell. The problem is easily dealt with, but advice from a vet is advisable.

Egg bound

A rare condition caused by an egg that is too big to pass. Symptoms include frequent visits to the nest box with no eggs being laid; a bird will also seem irritated, and can quickly become ill looking and go off her food. If the egg is visible, smear petroleum jelly over the vent to help lubricate it and hold her over a bowl of boiling water so the steam relaxes her muscles, though not too close as you could scold her. If this fails and you can see the egg, very carefully make a hole in it and remove all of it with your finger. Any that is left can cause an infection and kill your chicken.

Eye infections

These are often a symptom of an underlying problem, so seek the advice of a vet. If an infection is evident over-the-counter eye drops such as Broline, available from chemists, can work very well. In any event, clean the eye of pus using cotton wool soaked in cooled boiled salt water at least three times a day, until the infection has cleared and for a few days after.

Feather pecking

This usually occurs because of boredom, though sometimes it can be due to lack of protein in a chicken's diet or prolonged hot weather. Before jumping to any conclusions, check that loss of feathers isn't because of the moult, or because a chicken has lice or mites. If, however, there are 'V' shaped pieces of feather missing, these are beak marks and your chicken is definitely being feather pecked. Breaking the habit can be a hard job, but it can be done. If the problem is lack of space, either reduce your flock or build a bigger run. Also add toys and distractions. If blood is shed, you need to isolate the victim and deal with her injuries. There are anti-peck sprays on the market, which are made to taste and smell unpleasant. I prefer Stockholm tar, as used on horses' hooves. Heat it slightly and apply a layer to the bald patches. It is very messy and your girls certainly won't look their best, but it will grow out.

THE CHICKEN MEDICINE CHEST

- Apple cider vinegar
- Purple spray/iodine
- Surgical spirit
- Petroleum jelly
- Tonic/poultry spice
- Garlic
- Cod liver oil
- Red mite treatment
- Wormer
- Louse treatment

For the offending chicken you can purchase a plastic bit that makes it impossible for her to close her beak completely, but allows her to still eat and drink. These can usually be removed after a couple of weeks. If you have plenty of free time, you can spray the offender with water every time you see her being naughty. Or you can introduce a mature cockerel to your flock, he will sort out the argument. If you think the problem is being caused by lack of protein, add mealworm or cod liver oil to their feed for a couple of weeks. This is also beneficial, as it will help the girls to re-grow their feathers.

Frost bite
Chickens are very tolerant of cold weather, provided their house is dry and draught free. Cocks are more likely to suffer than hens as they do not tuck their heads in under their wings at night and their combs and wattles are larger. Smearing petroleum jelly on these regularly will help prevent this.

Heat stroke
This can put chickens off their feed and result in them stopping laying, laying less or laying smaller eggs. Chickens cannot sweat, but instead pant to cool themselves. Heavy breeds suffer more from heat than lighter breeds. White birds are less susceptible as they reflect heat better than dark birds. During hot weather providing shade and plenty of cool water is essential. I half fill bottles with water, freeze them, and then place in the run. Rehydration salts can also help. Hosing down the outside of their house can help cool temperatures, and misting adult birds with a fine spray can make them more comfortable.

Internal laying
This is due to the wave-like motion that moves the egg down the oviduct working in reverse and is very rare. Eggs are laid internally into the abdomen. As this fills with eggs, the chicken's rear end will lower because of the weight and she will begin to walk like a penguin. No one knows why this happens and there is no known cure, so humane despatch will be required.

Lost nails
Chickens occasionally lose their nails. If they do, a good clean, followed by an antiseptic spray, is all that is necessary.

Marek's Disease
A highly contagious viral disease that affects mainly young birds, for which there is no cure. Infected birds will look extremely ill and have paralysis of the legs and wings. Silkies and Sebrights are the two most susceptible pure breeds, however most hybrids are vaccinated as day-old chicks.

ABOVE: Feathers worn away by the treading of an amorous cock.
RIGHT: A group of pure breed pullets eagerly coming out in the morning.

Pasting

Occurs when loose droppings stick to the vent area of chicks. It is thought to be caused by inadequate diet and/or the chick becoming too cold. It does not spread from chick to chick. Unless you pick the droppings away, the vent may get sealed shut and the chick will die. Adding small quantities of grass or finely cut up lettuce to their feed can help. If it does not clear you will need to cull the affected chick.

Sour crop

The first sign of this is an enlarged crop that feels like a balloon part-filled with water. The usual cause is an unbalanced diet (too many scraps of food) and/or mouldy layer's pellets. To clear the crop, hang the bird by her legs and gently massage the liquid out. You will need to do this a couple of times but not for too long as it will be uncomfortable for her. Once the crop is cleared, add probiotic yogurt to her layer's pellets for a few days; this helps balance the micro-flora in her digestive system

Splayed legs

This happens when chicks are raised on a slippery surface or if the incubator temperature was too high. If a chick does hatch with splayed legs, it is possible to 'hobble' them using a small strip of plaster to tape their legs into the correct position, though only for a few days. If caught early, this should correct the problem. If it doesn't, the chick will need to be humanely culled.

Sprains and broken legs

A chicken's legs can often suffer from sprains and require rest. Causes include perches that are too high and over-amorous cockerels. Heavy, overweight birds are also prone to this problem. Broken legs are very rare and humane despatch is the best solution. However, I had a customer who had his hen's leg amputated by a vet because he had accidentally trodden on it himself and broken it. Although she adapted well to her new life, she obviously couldn't scratch around, which is a chicken's favourite pastime, and was kept on her own because the other girls bullied her. In my opinion it is cruel to keep a chicken on its own as they are flock animals, so when thinking of treatments, please think of the chicken and not yourselves.

Tumours

Tumours are most likely to form internally on the reproductive organs. They are a common cause for poor egg laying. Some tumours are an indication of other problems such as Marek's Disease, but others are not. Slow-growing tumours in older hens are common but little is known about them.

Death

Despatch is a nicer way of saying kill, which makes nearly everyone gasp. Unfortunately it is a part of keeping chickens that you do need to know about.

Despatching a chicken can be done by your vet if you really cannot face it, but if your girl is suffering, making her wait for an appointment and the stress of the journey can be extremely unfair. For many years I never had to despatch any as luckily my husband was always on hand to do the deed. Then he decided they were my chickens and my responsibility, and refused to do it again. I was shocked but as he could never remember my birthday, I thought he wouldn't remember his decision. Months passed, then Gertie, my eight-year-old hen, lost the use of her legs. She was in a sorry state and there was only one kind thing to do for her. I shouted for my husband, who smiled and said, 'Your chickens, your responsibility.' Feeling furious but determined I scooped her up and cuddled her, and took her to a quiet place away from the other girls. There was the log and the axe we'd used for the Christmas turkey (I named the turkey Trevor after my husband to make it easier to despatch) but I couldn't use it, so, with tears running down my cheeks, I dislocated her neck. It was quick and remarkably easy, apart from the emotional side, which I will never find easy. But I'm glad my husband made me; otherwise I would have never done it.

Despatching devices that work by crushing the neck and spinal cord rather than dislocating it can be purchased from the internet. However, the Humane Slaughter Association (HSA) does not recommend these. Where possible, they advise using electrical or mechanical concussion stunning first, followed immediately by a killing method such as neck dislocation, chopping or throat slitting. Hand-held electrical stunners and mechanical percussive devices are available, but they are expensive.

Neck dislocation is a quick and easy way of despatching, as long as you know what you are doing. Never hesitate, and always do it smoothly and quickly. First catch your girl calmly, talk to her, and take her to a quiet place away from the other girls. Then, hold the legs in your left hand if you are right-handed, or the right hand if you are left-handed. With your dominant hand place the head in between your first and second finger and your thumb under the beak, tilting her head slightly. With a very firm, meaningful, downward action pull the neck and bring the head backward. You should feel a pop as the neck breaks. When you hear this the job is done. She will flap and kick for a short whole. This is perfectly normal as it is only the reaction of the nervous system.

What can go wrong? If you do not pull down hard and firmly you will end up stretching the neck and not actually killing her, so if you are going to attempt it do not hesitate or do it half-heartedly. The worst-case scenario of pulling too hard is that her head will come off, which is very unpleasant for you but at least you did the deed thoroughly.

Chopping the head off sounds horrendous, but is very effective and little can go wrong. You need a large, flat log with two large nails nailed into it about 4cm apart and a very sharp axe. Calmly take your girl to an area away from the other girls. Place her head between the nails and squeeze the nails together to hold the head in place. In one firm movement chop off the head with the axe.

Throat slitting is best done using a killing cone, which is placed over the chicken's head so that her head pops out of the bottom. You then take a very sharp knife and cut her throat, making sure you cut the jugular vein. She will bleed to death very quickly. This method is favoured by smallholders who raise chickens for meat.

Normally chickens live for between six and eight years. The oldest chicken, according to the The Guinness Book of Records lived until the ripe old age of 16.

ABOVE: A chicken enjoying an afternoon wander.

Glossary

Air cell the space at the broad end of the egg. Its size indicates the freshness of the egg and, during incubation, the humidity levels.

Bantam miniature breeds, about a quarter the size of normal chickens. See also True bantam.

Barring striped pattern on feathers of light and dark colouring alternating horizontally.

Beard the head feathers beneath the beak.

Bedding material laid on the floor of a coop for comfort, insulation and for soaking up faeces.

Bloom protective layer deposited on an egg at laying.

Broody the stage in a chicken's reproductive cycle during which it incubates eggs.

Bumble foot a swelling of the foot.

Candling a method for seeing inside an egg by holding it up to a beam of light to illuminate its contents.

Capon a castrated cock fattened for meat.

Chick crumb chicken feed with a high protein content designed for chicks aged 0–6/8 weeks.

Chook Australian slang for a chicken.

Clipping cutting through the flight feathers of a single wing to stop the bird from flying.

Cloaca another term for a chicken's bottom. *See also* Vent.

Cock a male chicken older than 12 months.

Cockerel male chicken younger than 12 months.

Comb the red membrane on a chicken's head. It indicates fertility and is often a target for attack.

Coop a pen or cage for containing poultry.

Crest the display of feathers on top of a chicken's head.

Crop the internal pouch at the base of a chicken's neck. It is the first sac of the digestive system and contains grit to grind food.

Cuckoo stripes on feathers, stripes that are not regular and may blend into each other.

Culling the humane killing of ill or unwanted birds.

De-beaking the removal of the tip of the upper beak, to prevent feather pecking.

Double lacing on feathers, where there are two areas of dark colour. *See also* Lacing.

Down soft, fluffy feathers that have no shaft; typical of chicks.

Drinker the container in which water is placed.

Dust bath a clean dry area of soil or sand where birds 'bathe'.

Egg tooth the beak of a chick; it is designed to crack the egg and is often referred to as the egg tooth.

Feeder the container in which feed is placed.

Frizzle feathers that curl backwards, giving a 'permed' appearance.

Gizzard the muscular pouch within the bird's digestive system, which helps grind their food.

Grit chickens need a supply of grit for grinding food in the crop.

Grit station the container in which grit is placed.

Grower's pellets ready-prepared feed that is suitable for growing chickens aged 6/8–16 weeks.

Hackle the feathered area where the breast meets the neck.

Harden off to gradually take away an artificial heat source from chicks.

Hen a female chicken over one year old, normally past its first annual moult.

Hopper another term for a feeder.

Hybrid a bird whose parentage comes from two distinctly different breeds.

In lay the term used to describe a chicken that's laying eggs.

Lacing where the edge of the feathers is a different colour to the middle. *See also* Double lacing.

Layer a chicken that is producing eggs.

Layer's pellets ready-prepared feed that is suitable for chickens laying eggs.

Mash ready-prepared chicken feed which is similar to pellets but in milled form.

Moult the method by which birds replace their feathers once a year.

Muff the feathers on a chicken's cheeks, attached to the beard. Also called whiskers.

Nest box the box in which hens lay. It should be positioned somewhere quiet and secluded.

Pecking order the order of dominance by which chickens arrange themselves.

Pellets ready-prepared chicken feed in bite-sized pieces which is usually a complete diet ration.

Pencilling very fine stripes which follow the shape of the feather.

Perch a rod or bar on which hens sleep above the ground, as though in a tree.

Pipping the way a chick breaks into the air sac in its egg before hatching.

Plumage a bird's complete coat of feathers.

Point-of-lay (P.O.L.) the stage of a chicken's life at which it can start to lay eggs. It can occur at any time between 16–24 weeks.

Pop hole entrance into a hen house or run, usually closed by a door or sliding shutter.

Poultry saddle a cloth placed on the back of a chicken to protect her from the male's spurs and claws or from feather pecking.

Preening the way in which a bird cleans and arranges its feathers.

Pullet a female chicken under one year old.

Pure breed hens that hatch male and female chicks that share common characteristics and always breed true.

Rearer's pellets *see* Grower's pellets.

Saddle the back area of a chicken, around the shoulders.

Scales the overlapping plates of skin on a chicken's legs.

Spangling feathers with a splodge of colour on the tip of each feather.

Spur the horny growth on the rear of a male's legs.

Treading an alternative term for mating.

True bantam a breed about a quarter the size of normal chickens but which has no large equivalent.

Vent a chicken's bottom. There is no vaginal opening in birds, so their eggs are laid through the same one as they excrete through.

Wattle the flesh hanging from the face by the beak. It is often red, like the comb.

Whiskers the feathers on a chicken's cheeks, attached to the beard. Also called muff.

Index

Index

Resources

CLOTHING

The Farmer's Friend
Suppliers of quality outdoor clothes. Fantastic, with a speedy service.
Telephone: 01392 277024
www.farmersfriend.co.uk

FEED

Dodson and Horrell Ltd
Supply top quality feed.
Telephone: 01270 782236
www.dodsonsand horrell.com

Fancy Feed Company
Speciality feed using non-GM ingredients.
Telephone: 01371 850247
www.fancyfeed company.co.uk

Marriages
Quality, traditional feed..
Telephone: 01245 612019
www.marriages.co.uk

Organic Feed Company
A range of 100 per cent organic, high quality feed. They supply River Cottage.
Telephone: 01362 822903
www.organicfeed.co.uk

Small Holder Range
Feed designed for animals being raised more naturally, made using GM-free ingredients without artificial growth promoters or artificial yolk pigmenters. Approved by the Vegetarian Society.
Telephone: 01362 822902
www.smallholderfeed.co.uk

HOUSES AND FENCING

Animal Arks
Supply a full range of recycled plastic products, all made to last forever, in a range of colours.
Telephone: 01579 382743
www.animalarks.co.uk

Flyte So Fancy
Design and hand-craft a range of poultry houses, including chicken runs, houses and coops. They also supply electric fencing.
Telephone: 01300 345229
www.flytesofancy.co.uk

Little Acre Products
Great coops and runs.
Telephone: 01827 215697
www.littleacre products.co.uk

Omlet
Modern, plastic chicken coops that are easy to clean and move.
Telephone: 01295 500900
www.omlet.co.uk

Smiths Sectional
Quality coops made by a small family company, who will also deliver and erect.
Telephone: 01630 673747
www.smithssectional buildings.co.uk

Fishers Woocrafts
Coops that are made to last and provide an excellent service.
Telephone: 01757 289786
www.fisherswood crafts.co.uk

Electric Fencing Direct
Provide a great service and good aftercare. Recommended.
Telephone: 01620 860058
www.electricfencing.co.uk

Kiwi Fencing
Supply both permanent and temporary fencing.
Telephone: 01728 688005
www.kiwifencing.co.uk

TONICS AND WORMER

Verm-X
Wormers and tonics that promote well-being.
Telephone: 01984 629125
www.verm-x.com

Flubenvet
A poultry wormer that is widely available.
Telephone: 01582 842096
www.viovet.co.uk

Life-Guard
A natural supplement for chickens with a patented formula of antioxidants and vitamins.
Telephone: 01582 842096
www.viovet.co.uk

Acknowledgements

Special thanks to all my friends for their help and support: Amanda, Ian, Roy and Judy for their celebrating in style! Ben and Cathie for just being them, absolute rocks. Ed, who manages to put up with me most days. Will for his humour and honest insults! Luke for all his knowledge and support when things go wrong. Everyone at Kyle Books who has made this possible. Lastly my four children and husband for their constant help and enthusiasm even in the snow, and their love.

Dedication
To Mum and Dad, for my fantastic childhood and where it has led!

———————————————————————

An Hachette UK Company
www.hachette.co.uk

First published as *Chickens: The essential guide to choosing and keeping happy, healthy hens* in Great Britain in 2012.

This edition *Raising Chickens: The essential guide to choosing and keeping happy, healthy hens* published in 2022 by
Kyle Books, an imprint of Octopus Publishing Group Limited
Carmelite House
50 Victoria Embankment
London EC4Y 0DZ
www.kylebooks.co.uk

ISBN: 978 1 91423 972 4

Distributed in the US by Hachette Book Group, 1290 Avenue of the Americas, 4th and 5th Floors, New York, NY 10104

Distributed in Canada by Canadian Manda Group, 664 Annette St., Toronto, Ontario, Canada M6S 2C8

Publishing Director: Judith Hannam
Publisher: Joanna Copestick
Editorial Assistant: Emma Hanson
Design: Laura Woussen
Photography: Christian Barnett
Copy editor: Liz Murray
Proofreader: Catherine Ward
Production: Lisa Pinnell

Printed and bound in China

10 9 8 7 6 5 4 3 2 1